"Some men see things as they are and say, 'Why?'
I dream of things that never were and say, 'Why not?'"

ROBERT F. KENNEDY,
1968 PRESIDENTIAL CAMPAIGN

RACE TO
MARS

DANA BERRY

A BARRON'S / MADISON PRESS BOOK

First Edition for the United States, its territories and dependencies, the Philippines, and Canada
published in 2007 by Barron's Educational Series, Inc.

This book is based on

the docudrama *Race To Mars* and the documentary *Mars Rising* produced by Galafilm with the financial
participation of Discovery Channel Canada, the Canadian Television Fund, the Quebec Film and Television
Tax Credit, the Canada Film or Video Production Tax Credit, 13 Production, ARTE France, Discovery
Communications, LLC (*Race to Mars* only), NHK, and Galafilm Distribution Inc.

and

the interactive *Race to Mars* project produced by QuickPlay Media Inc. with the financial participation of the
Quebecor Fund, the Bell Broadcast and New Media Fund, Telefilm Canada Administrator of the Canada
New Media Fund funded by the department of Canadian Heritage, and the Ontario Media Development
Corporation's Interactive Digital Media Fund.

All inquiries should be addressed to:
Barron's Educational Series, Inc.
250 Wireless Boulevard
Hauppauge, NY 11788
www.barronseduc.com

Library of Congress Control Number: 2005921556

ISBN-13: 978-0-7641-5905-3
ISBN-10: 0-7641-5905-4

Produced by
Madison Press Books
1000 Yonge Street, Suite 200
Toronto, Ontario
Canada M4W 2K2

Printed in China
987654321

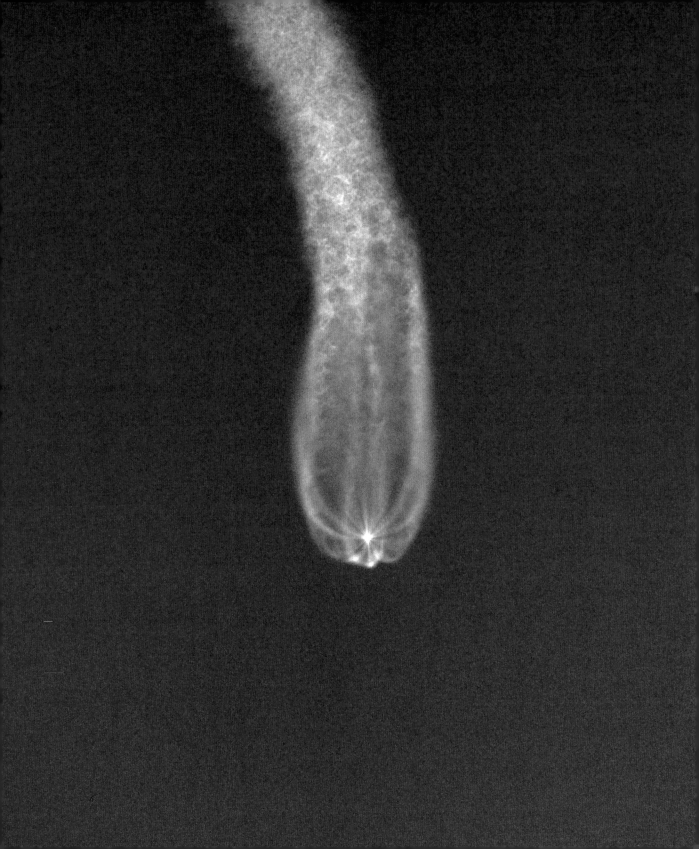

CONTENTS

CHARTING THE JOURNEY — 8

CHAPTER ONE
MARS 2030 — 10

CHAPTER TWO
PYGMALION'S PLANET — 30

CHAPTER THREE
MARS IN THE SPACE AGE — 62

CHAPTER FOUR
GETTING THERE — 90

CHAPTER FIVE
COME IN, MARS — 124

CHAPTER SIX
LIFE ON MARS — 154

CHAPTER SEVEN
GOING HOME — 178

ACKNOWLEDGMENTS, CREDITS AND INDEX — 184

CHARTING THE JOURNEY

Planning an expedition to Mars, like planning any trip, begins by choosing the best route. Because Earth and Mars are in constant motion around the Sun, pathways to Mars come under two general categories: *Opposition class*, when Mars and Earth are close by; and *conjunction class*, when Mars and Earth are on opposite sides of the Sun.

The voyage to Mars described in this book follows an opposition class route; it takes advantage of the gravitational field of Venus to accelerate the spacecraft on its outbound journey. This also reduces the amount of fuel needed for propulsion, and affords humans their first close-up view of Venus as they swing by. The downside of this route is that it almost doubles the length of the outbound voyage — to eleven months.

Following the chart at right, *Terra Nova* will leave Earth with a crew of six from position number 1, on January 30, 2030. On July 4, *Terra Nova* will make its closest approach to Venus at position 2 on the chart and will receive a gravitational boost during the flyby. On July 20, there will be an unplanned encounter with a previously uncharted near-Earth asteroid.

At Position 3, *Terra Nova* will be inserted into Mars orbit and will dock with the *Gagarin* Mars ascent/descent vehicle. Humans will touch down on Mars starting December 23, 2030, just as the planet approaches opposition with Earth. The duration of this Mars excursion phase is marked on the chart along the Martian orbit in red and lasts until February 21, 2031. At position number 4, humans will return to *Terra Nova* via *Gagarin* and leave Mars. At position 5, *Terra Nova* catches up with Earth and the crew returns safely home.

Jupiter

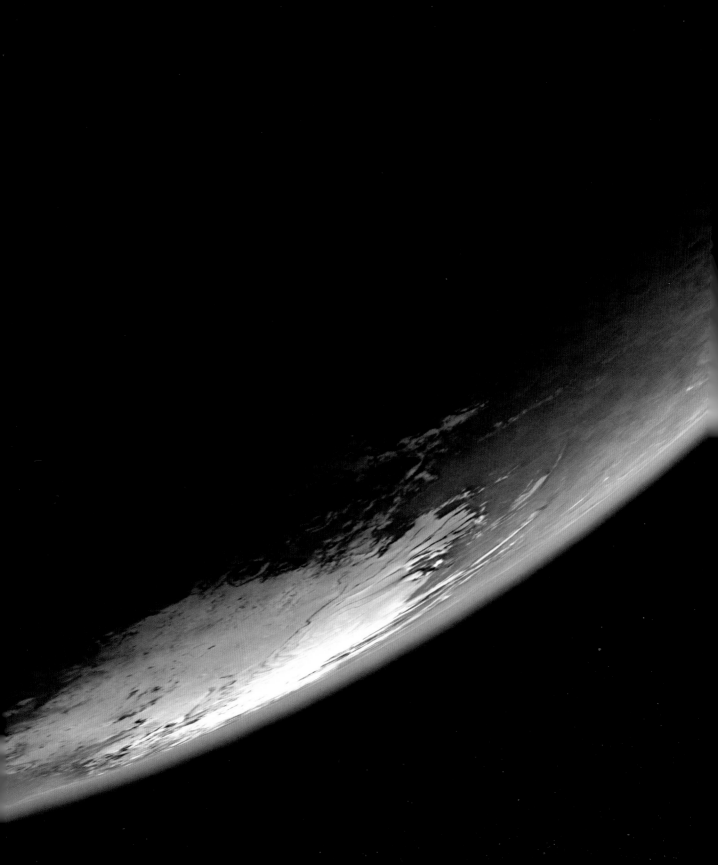

MARS 2030

"MEN WANTED FOR HAZARDOUS JOURNEY. SMALL WAGES, BITTER COLD, LONG MONTHS OF COMPLETE DARKNESS, CONSTANT DANGER, SAFE RETURN DOUBTFUL. HONOR AND RECOGNITION IN CASE OF SUCCESS."
— Ernest Henry Shackleton, Antarctic explorer, 1914

October 22, 1968. My dad and I are watching live coverage of the *Apollo 7* splashdown. The television announcer says the returning capsule will be visible to anyone living in the southeastern United States as it crosses Florida on its way to the target landing zone in the Atlantic Ocean. My dad and I rush outside just in time to see an orange streak moving from west to east across the morning sky. We gaze in awestruck silence as it slips below the horizon and is gone.

As *Terra Nova* approaches Mars, the south polar ice cap comes into view.

Shortly after we saw it, *Apollo 7* successfully splashed down somewhere southeast of Bermuda. It was the only *Apollo* capsule to come down in the Atlantic; all the others were in the Pacific. So that crisp October morning was the only chance that most Americans had to see with their own eyes an *Apollo* spacecraft coming home.

I had studied the history of exploration in school. I knew all about the wooden ships that had once sailed the oceans in search of new worlds. I dreamed that in my lifetime we would explore new worlds in rockets. Seeing *Apollo 7* with my own eyes made that dream more real than it had ever been before.

After the Moon landings, everyone knew that Mars would be next. Our fascination with Mars has had a long and curious history. No human has been to Mars, although it seems as if a parade of satellites and robot missions embarks for the Red Planet every year. These emissaries have dispelled some ancient misconceptions about the planet, and they have given us a peek at what will await travelers when they arrive from Earth. But seeing pictures of a landscape through the eyes of a robot is like watching a space capsule return to Earth on television. It can't compare to the excitement of seeing the capsule itself. And no disembodied mission to Mars has the same visceral significance as the knowledge that a human has walked on the Red Planet.

It will happen one day. The plans are being drawn up now. In these pages we imagine how that first human mission to Mars might be accomplished.

The first manned flight in the *Apollo* program draws to a successful end as the *Apollo 7* command module descends to the surface of the Atlantic Ocean. The 10-day mission was the longest spaceflight up to that time, longer than all Russian space missions combined.

GOING IN

December 22, 2030. *Terra Nova* is by far the largest spacecraft ever to attempt orbital insertion around the Red Planet, and the first to carry passengers there. The men and women chosen for this journey excel in every aspect of their specialties and training, but on this day they are excited, perhaps even anxious: they know better than anyone that a miscalculation now would probably be fatal. The pilot-commander is ready to press the manual override switch at the first hint of trouble.

Back at Mission Control at the Johnson Space Center in Houston, Texas, engineers sit at their consoles, tense and motionless in the flickering electronic light. All eyes are on the enormous monitor above them. The video feed is breathtaking. There on the screen is a view from one of four external crew-cabin cameras — the horizon of Mars fills the bottom portion of the picture, while the Mylar-wrapped truss of *Terra Nova* protrudes across the top. No chronicle of exploration has ever described a journey like this. Never has there been an enterprise so laden with opportunity, vision, and hope.

The list of superlatives generated by the flight of *Terra Nova* seems almost as long as the journey to Mars itself. It is the largest manned spacecraft ever to leave Earth orbit. It carries the first humans to use a gravity boost by slingshotting past another planet, the first to fly past Venus, the first to arrive over Mars, and the first to behold a tiny, pale-blue crescent called Earth.

Terra Nova embodies many technological breakthroughs. The nuclear thermal rocket (NTR) is the product of a decades-long process of invention and improvement. The base camp and its compact nuclear power plant, all assembled and waiting for the astronauts' arrival on Mars, represent an exciting new approach to space exploration.

It isn't a venture for the faint of heart.

THE ROCKET

The nuclear thermal rocket (NTR) engine was originally designed for the Prometheus test bed probe to Jupiter by Stanley Borowski at the NASA Glenn Research Center in Ohio. It was subsequently modified by the Advanced Propulsion Lab at NASA's Jet Propulsion Lab in Pasadena, California. Borowski's design was similar to the Nuclear Engine for Rocket Vehicle Applications (NERVA) project that was originally proposed for use as an upper-stage rocket on the *Saturn V Apollo* moon rocket.

NERVA worked by forcing hydrogen gas among the hot coils of graphite rods embedded with uranium carbide (UC). The heated hydrogen is then expanded through a nozzle to produce thrust. When NERVA was first tested in the 1960s, it produced thrust of over 250,000 pounds (113,000 kilograms). *Terra Nova*'s NTR uses uranium oxide (UO) in a tungsten (W) metal rod. Thrust is enhanced by mixing oxygen into the nozzle as the superheated hydrogen gas escapes. This causes an afterburner effect that enhances thrust by some 20 percent.

2030:12:22

HOME SWEET HOME: The 325-ton (295-metric ton) *Terra Nova* is the world's first manned interplanetary spacecraft, a twenty-first-century caravel for the exploration of the high frontiers of space. In the foreground on the right is the Trans Hab, the multi-deck module containing the crew quarters, wardroom, airlock, sick bay, science lab, flight controls, and all supplies and consumables. Behind the Trans Hab are four fuel tanks attached to a 400-foot (120-meter) truss. Also attached to this truss are the nuclear rocket motor and the Earth return capsule. Two launches are required to assemble *Terra Nova* in low Earth orbit.

Before the orbital insertion burn of its NTR, *Terra Nova* first has to spin down. It has been rotating like a windmill propeller throughout the months-long voyage to create artificial gravity through centrifugal force. For the crew, the termination of artificial gravity brings on weightlessness and the queasy feeling that weightlessness induces. In this instance, the queasiness is amplified by knowledge of the dangers that still lie ahead. Historically, robotic missions achieved capture into Martian orbit by skimming through the upper layers of the Martian atmosphere — a technique called *aerobraking*. A spacecraft takes a momentary dip into the Martian air, where friction from the atmosphere slows the spacecraft down. The advantage of this technique is that it reduces the amount of fuel the spacecraft must carry. But there are risks as well: hit the atmosphere at too shallow an angle and the spacecraft may skip off into space, like a stone skipping across water. Come in too steeply and the spacecraft will burn like a meteor. The history of Mars exploration is littered with the charred wreckage of robot spacecraft that came too close or entered at the wrong angle.

One of the major benefits of NTR propulsion is that it removes the need to aerobrake, but capture remains a delicate maneuver. The crew will fire the NTR engines for a propulsive capture into Mars orbit and then transfer from *Terra Nova* to *Gagarin*, the Mars ascent/descent vehicle (MADV), in order to descend to the Martian surface. To compound the risk, solar max has been coming early; normally, solar max is a time when the number of sunspots on the Sun's surface reaches its peak in an approximately 11-year cycle. (The last solar max as of this writing was 2001.) The number of sunspots corresponds with the intensity of solar radiation: The greater the radiation from the Sun, the more the Martian atmosphere expands. The year 2030 was chosen specifically because although it is not a solar max year, there has been an unusually high amount of sunspotting. As a result, there is some uncertainty about whether or not *Terra Nova* will brush the atmosphere.

So the adrenaline is pumping as the six members of *Terra Nova*'s crew strap themselves into their seats and brace for the last NTR burn of their outbound mission. If the final course correction is off by just a thousandth of a degree, or the last engine burn terminates a few seconds too soon, their insertion into orbit will fail. But there is nothing they can do about it now. They are going in.

There is a sudden bang, and *Terra Nova* begins to shake. It is reminiscent of a scene from a twentieth-century movie, *Contact*, in which Jodi Foster's character Ellie goes through the wormhole, or an old war movie in which submarines are pounded from the outside by depth charges. The rattling grows, and there are loud, mysterious pops and crashes. Has something come loose? Is the

Gagarin Mars ascent/descent vehicle (named for the first human in space) is pre-deployed in low Mars orbit ahead of *Terra Nova*. When the two spacecraft rendezvous, *Gagarin* is used to transfer the crew to and from the Martian surface.

ship cracking up? Have they miscalculated? Space travelers do not relish surprises and this bumpiness is a surprise. A long, loud groan, as if the ship itself is sighing, passes through the spacecraft. Is this the end? If the ship is going in at too steep an angle, at least it will be over quickly and there will be no suffering. And then, just as suddenly as it began, the rattling stops.

MARS WANTS YOU

Think you've got the right stuff to go to Mars? Astronaut recruitment at National Aeronautics and Space Administration (NASA) is ongoing and is subdivided into three categories: pilot, mission specialist, and payload specialist. All three positions require a college degree, and the mission and payload specialists both must have a degree beyond a B.A. or B.Sc. The pilot's job also requires a minimum of 1,000 hours of flight time in jets, as well as strong leadership skills. Individual aptitude, motivation, sensitivity, and emotional maturity are all criteria by which an applicant to the astronaut corps is judged, and these form the basis for the selection of a crew bound for Mars.

Would-be Mars travelers face a variety of psychological and sociological challenges. They must cope with long periods of isolation and confinement that could lead to irritability, loneliness, pessimism, and depression. Does this make a lone-wolf personality the ideal personality for a Mars astronaut? Not necessarily. Although loners like Kurt Cobain and James Dean are celebrated in popular culture, their "live fast, die young" credo has no place in space. A loner would have to curb his or her antisocial tendencies before embarking for Mars. This is not impossible: American astronaut Neil Armstrong was a loner by temperament, but he made the adjustment.

Neil Armstrong in an early 1969 training session for the *Apollo 11* moon landing.

Group dynamics are also a factor in the selection process. Does the individual work well with the team? Does everyone contribute equally, and can they overcome cultural and gender issues? How well would the group handle trauma? How would they face the death of one of their own, or the death of a family member back on Earth? Can the group avoid marginalizing and scapegoating one of its members? A single mistake by any member of the expedition could lead to catastrophe. This knowledge could weigh heavily both on individuals and on the group, leading to intolerable tension if solid relationships are not established first.

There have been a variety of analogous, similarly fraught group endeavors: scientific research conducted in the Antarctic, the lengthy undersea voyages of nuclear submarines, and, of course, previous long-duration space missions aboard *Mir* and the *International Space Station (ISS)* provide a few examples. But, in the final analysis, the only way to understand the challenges of an unprecedented trip is to take the trip.

On this, the first interplanetary voyage, six astronaut candidates have been chosen for the *Terra Nova* crew: two NASA astronauts, one Russian cosmonaut, one Canadian Space Agency (CSA) astronaut, one European Space Agency (ESA) astronaut, and one Japanese Aerospace Exploration Agency (JAXA) astronaut.

Every one of them is a veteran space traveler, having logged months at the *International Space Station* and on the Moon. They have spent years training together at the facilities of each participant's home space agency. The idea behind this preparation is to provide each member of the crew with an understanding of the work methods and cultural backgrounds of the different agencies. It's vital that they all get along.

In reaching for the stars, this first interplanetary space mission may also accomplish something else — creation of a close understanding among the men and women of separate nations who have first reached out to one another before embarking on a great and dangerous task.

LANDING ON MARS

The Mars ascent/descent vehicle *Gagarin* was en route to Mars before *Terra Nova* even left Earth. For the crew on board *Terra Nova*, the next stage of the mission is the rendezvous with the MADV. This is another tricky maneuver, made even trickier because, for the crew, it comes after roughly 11 months of space-flight. They're tired, and their skills are rusty. Such fatigue played a major role in the collision of the *Progress Supply Module* with *Mir* on June 25, 1997. *Terra Nova*'s crew is neither as fresh nor as relaxed as were the cosmonauts on *Mir*. But, after some careful ad hoc maneuvers, *Gagarin* is captured with very little fuss.

As luck would have it, just at this time a large dust storm begins spreading across the Martian surface. In a little less than 24 hours the storm will reach the Dao Valles landing site, so a decision is made to attempt the landing a day early. All six astronauts board *Gagarin* and strap themselves down in preparation for the shock of entering the Martian atmosphere.

The meticulously constructed set built for the film, *Race to Mars*, reflects both the grandeur and the desolation of the landing site in Dao Valles.

RED PLANET RENDEZVOUS: With Mars as a backdrop, this sequence of three inset images shows *Terra Nova* capturing *Gagarin* in low Mars orbit. On the left, the two spacecraft have acquired a radar lock on each other and are positioning themselves for the docking. In the center picture, the spacecraft are about to make contact. This is the last chance to go to manual override and abort the docking sequence if anything is judged to be out of sync. In the third picture, the two spacecraft have completed the first orbital rendezvous to take place over another world.

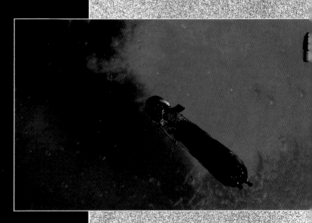

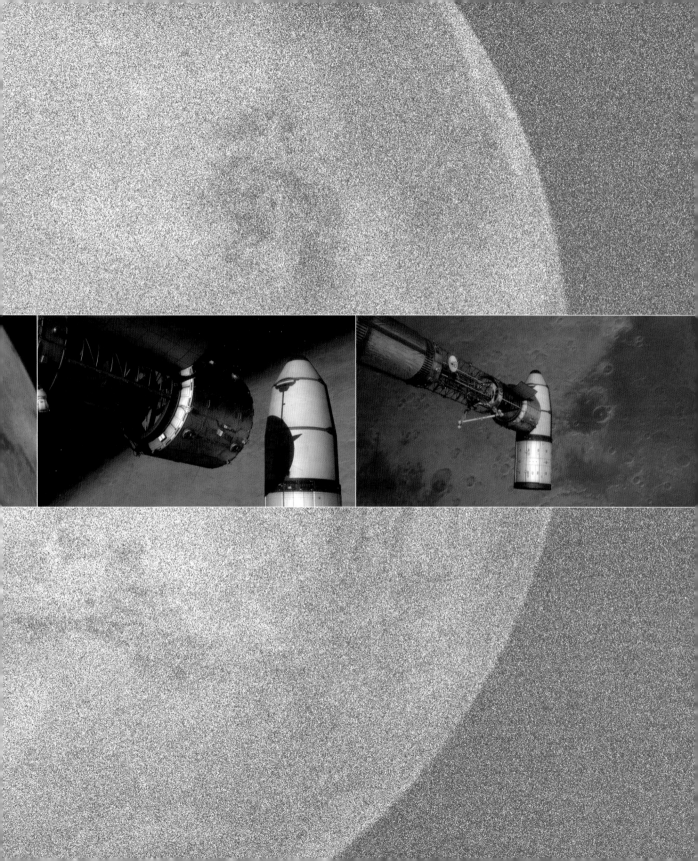

With a thump *Gagarin* is released and pushed away from *Terra Nova* by springs embedded in the docking ring. As *Gagarin* backs away, it rolls over 180 degrees, affording the crew a view of the massive rift valley, Valles Marineris. *Gagarin* will attempt to land within yards of the base camp, where the *Atlantis* habitat module and all the rest of the equipment have been set up. The coordinates are longitude 266 degrees west, latitude 33 degrees south, at the source for the Dao Valles canyon complex in the northeast corner of the Hellas Basin. This location has been chosen in part because the canyon floor is generally smooth, and in part for its proximity to the Hadriaca Patera volcano, with its diverse geological makeup.

The pilot and commander monitor the landing sequence as they enter the Martian atmosphere. Although *Gagarin* is a pressurized spacecraft, the crew is suited up in case there is a mishap leading to a crash landing.

But first the crew has to get to the planet's surface. Outside their spacecraft, the sky is changing from the star-filled black they have known for the past 11 months to an enveloping pale red. As *Gagarin* streaks across the sky, small, drogue parachutes open up. These, in turn, release larger parachutes. With a slight bump the two halves of the aero-descent heat shell are blown off, revealing the nozzle of the main descent engine. The parachutes are released as the engine roars to life and burns for the remainder of the descent. Retro rockets puff sporadically and in rapid, computer-controlled succession, guiding and adjusting the lander as it makes its way down.

TOP: The heat shield on *Gagarin* glows as friction builds up. *Gagarin* hits the Martian atmosphere at a speed of Mach 12.6578.

BOTTOM: When it has slowed to Mach 3, drogue parachutes are released, which in turn pull out the main parachutes at an altitude of 5.6 miles (9 kilometers).

For months, the crew has been trapped inside their cramped, fetid spacecraft, dreaming of their destination. But now that their dream is about to come true, they are filled with apprehension. What will they find? Will they be equal to the challenge? Will the equipment still work? Will they ever go home again? If the engineers associated with the space program have learned one thing, it's that Murphy's Law is a real law of the universe. Experience with everything from the *Hubble Space Telescope* to the *International Space Station* has taught them that major systemic malfunctions can and do occur.

On Earth, the staff at Mission Control holds its collective breath, awaiting the outcome of the final maneuvers. The time lag in communication between Earth and Mars makes the suspense especially unnerving. Long minutes will have passed by the time they learn what has happened to the crew.

Back on *Gagarin*, there is a jolt as the shock absorbers in the spacecraft's legs absorb the force of the landing. The ship rocks a little and settles into the Martian landscape.

LEFT: Twenty seconds after the parachutes are deployed, *Gagarin* decelerates to Mach 2. The parachutes are then detached and the aeroshell is blown off as the landing rockets roar to life. A soft landing on Mars is only moments away.

OPPOSITE PAGE: This 2006 *Mars Reconnaissance Orbiter* view of Mawrth Vallis has been enhanced to accentuate color differences.

PYGMALION'S PLANET

As *Gagarin* settles into its new home at the Dao Valles base, it also settles into history. Its landing is the culmination of millennia of dreams and lore.

Our story begins with the Roman poet Ovid, who told us the tale of a young sculptor named Pygmalion who lived on the island of Cyprus. He was gangly and awkward, and this made him uncomfortable around women. In fact, he was so tongue-tied and self-conscious that he found himself increasingly reluctant to seek female company. It was so much easier to hide than to endure the mortification caused by his social ineptitude.

But what Pygmalion lacked in social skills he more than made up for in artistic talent. He was a superlative sculptor, and no one was better.

Gagarin makes history when it lands in the Dao Valles canyon complex on Mars.

And yet, he was lonely. So one day, Pygmalion began work on something new. He used large pieces of ivory and chipped and scraped and sanded and smoothed until he had fashioned the likeness of a maiden whose beauty was unmatched in the land.

When he stood back, Pygmalion was dazzled by the woman he had created. She was the embodiment of all of his dreams and desires. He fell in love with her, brought her flowers and rings, and adorned her with such trinkets as a woman might want.

As this was happening, the festival of Aphrodite was underway. The celebrations were magnificent, and many offerings were made to the goddess of love. When Pygmalion made his offerings, he also uttered a prayer. Aphrodite took pity on the young sculptor and smiled.

When Pygmalion returned home, he tenderly kissed his beloved statue, and to his astonishment, the statue was soft and warm, and it kissed him back. Aphrodite had answered Pygmalion's prayer and made from his statue Galatea, a sea nymph.

Galatea bore two children by Pygmalion, Paphus and Metharme, and in the classic fashion of all great love stories, they lived happily ever after. The End.

This story, which is a fusion of interpretations of Ovid by Thomas Bulfinch and Robert Graves, serves as an allegory for humankind's history with Mars. For the story of this relationship is not really a tale of battlefields and war; it is instead a love story. Its plot is driven by our need to find patterns or to impose them.

A panel from the Bayeux Tapestry records the panic caused by the appearance of Halley's Comet during the coronation of King Harold in 1066. The apprehension of the nobles pointing toward the comet is surpassed only by the wide-eyed terror of the king (right) as he learns of the arrival of this celestial interloper.

Throughout the course of history, we have seen Mars the way we see faces in clouds and behold a world vaguely familiar. The shining Red Planet has been seen as the embodiment of all of our dreams and desires, the alluring promise of understanding and fulfillment, and the gift of meaning. Mars is our creation and we have fallen in love.

THE PARLIAMENT OF STARS

Our Pygmalion-like fascination with Mars has roots that date back to our earliest awareness of the planets and the stars. Some stars appeared to move randomly, with a kind of capriciousness that was difficult to understand. The apparent randomness of comets led early observers to think of them as omens of misfortune. An image of perhaps the most famous comet was stitched into the fabric of the Bayeux Tapestry. This tapestry, woven in 1077, records the appearance in 1066 of a comet at the coronation of the English king Harold, successor to Edward the Confessor. The tapestry shows a procession of retainers and nobles distracted by the comet, which looms ominously over Harold's palace. Harold himself seems to be gripped with fear as he is told of the arrival of this celestial interloper.

This particular comet eventually came to be known as Halley's Comet, named for Sir Edmund Halley after he predicted in 1703 that it could be expected to reappear after an interval of 76 years.

It wasn't only comets that attracted attention. The apparently capricious behavior of planets also puzzled early observers. Planets seemed to migrate in one direction and then, for no apparent reason, reverse course, and then revert back to their original paths. This S-shaped motion applied to all the planets that lay beyond Earth's orbit around the Sun; those known to the ancients were Jupiter, Saturn, and Mars. This S or retrograde motion was later found to be caused by Earth overtaking and passing the outer planets as it circles the Sun. This happens because Earth swings around the Sun at a much faster pace than the outer planets. When Earth overtakes a planet, it has completed a *synodic period*, which is signified by the retrograde motion.

Of all the wandering stars, Mars provoked the most wonder. Was it the red hue, unique to Mars? Or that strange S motion? The Romans associated the Red Planet with war and violence and it is they who gave this planet its modern name, which is derived from that of their own god of war.

THE VIEW FROM THE ACROPOLIS

In ancient times, a common belief was that Earth occupied a fixed position at the center of the Universe, and that the stars circled around it as if each were attached to one or another of a series of nested spheres. This *geocentric* model of the Universe had many adherents, each of whom put forward a different version of the model. Aristotle, for example, claimed that the number of spheres encircling Earth was 55.

The publication in 1683 of Alain Manesson Mallet's 5-volume *Description de l'univers* featured one of the first pictures to record the actual features of the Red Planet. There is a temptation to discount such early renderings as primitive. However, the significance of the work does not rest with its topographical fidelity, but rather with its mere existence — Mallet shows us that Mars is a real place.

FIGURE XLVII.

Opposing the geocentric model was a competing *heliocentric* theory predicated on the notion that the Sun is in the middle of the cosmos. Astronomers of the Pythagorean school (none of the work of Pythagoras himself has survived) believed Earth is but one of several planets orbiting the Sun. Aristarchus was a proponent of this idea, and according to legend he wrote a proof that was later lost when the library of Alexandria was destroyed by fire.

The problem with the heliocentric model was that it did not seem to match our common experience. Earth seemed large, heavy, and fixed. It was unreasonable to suppose that our world with its mountains, valleys, and seas was swinging around the Sun.

In the second century C.E., the Greco-Roman astronomer Claudius Ptolemaeus (better known as Ptolemy) refined the geocentric model in *The Almagest*. This brilliant book distilled the accomplishments of 1,000 years of Babylonian, Chaldean, and Greek astronomy. The arguments set forth by Ptolemy were so persuasive that *The Almagest* laid to rest all competing theories and remained unchallenged for more than 1,300 years. Its preeminence was enhanced, in no small measure, when it was adopted as official Church dogma by the Catholic Pope Nicholas V, who commissioned a translation of *The Almagest* in the mid-fifteenth century.

THE REVOLUTION

The light of modern science began to shine when a Polish cleric, amateur astronomer, and neo-Platonist, Nicholas Copernicus, published a work describing what was to become a radical new cosmological model, a sixteenth-century version of the minimalist, less-is-more philosophy. His model of the cosmos simplified

Johannes Kepler (1571–1630) was a devout Lutheran who believed that the workings of the cosmos could be understood by rational thought and intellectual inquiry.

what Copernicus saw as a needlessly complicated Rube Goldberg contraption that made no sense. Like Plato, he sought a perfect underlying geometric order, and so he retained the nested crystalline spheres of the Ptolemaic model and the perfect circles they implied. However, these spheres were now heliocentric, that is, centered on the Sun as the Pythagorean school had suggested, rather than on Earth.

But Copernicus lived in dangerous times. He was frightened by the consequences that the publication of his ideas might bring. After all, the air was filled with the odor of burning witches. Still, one of his followers, an Austrian who used the alias Rheticus, persuaded Copernicus to publish his findings and began the preparatory work for publication.

Rheticus was unable to complete this task — we're not sure why — so he handed the job over to a friend. Andreas Osiander delivered the first edition of *De Revolutionibus Orbium Coelestium* in 1543. Not all was well, however. From his deathbed, Copernicus saw that the introduction he had written had been replaced by an anonymously written disavowal of the book's contents.

The author of this sabotage wasn't known until decades later when Johannes Kepler identified it as the work of Osiander. A Lutheran theologian, Osiander believed that truth was known only through divine revelation. Had he been confronted with his handiwork, Osiander probably would have claimed it was to shield Copernicus from the opprobrium of the Church. Copernicus had launched the first modern scientific revolution, and Osiander was its Benedict Arnold.

Copernicus passed away immediately after publication of his monumental work and thus was safely beyond the reach of religious zealots. Others engaged in astronomical observation were not as fortunate. Some, like Galileo Galilei, were kept under house arrest. Others fared much worse: Giordano Bruno, for example, was strapped upside-down to a stake and burned to death. He was executed in the year 1600 for advocating the notion that there were other stars with planets and that people lived on those planets. Worse still, Bruno said that the Copernican model was correct.

That same year, another nonconformist was about to change the world forever, and Mars was the key to that change. The Copernican model, whether one agreed with it or not, had re-opened the debate about the nature of the cosmos.

"A CARTFUL OF DUNG"

About 900 miles (1,500 kilometers) to the north of Rome there was once a great castle on the tiny island of Hven in the Oresund Strait. The Danish astronomer Tycho Brahe dedicated the castle to the muse of astronomy, Urania. He called it Uraniborg and fitted it out using funding from the king of Denmark, Frederick II.

Tycho Brahe (1546–1601) was the last and greatest of the pre-telescope astronomers. He knew that his naked-eye observations were the best in the world, but he also understood that he lacked the theoretical skills required to make sense of them.

> "By the study of the orbit of Mars, we must either arrive at the secrets of astronomy or forever remain in ignorance of them."
>
> — JOHANNES KEPLER

The facilities at Uraniborg included an observatory, a paper mill, a printing press, an alchemist's furnace, and a library, in addition to comfortable living quarters. It should have been a great place to conduct serious astronomical research, but there were distractions. Tycho threw outrageous parties attended by a who's who of princes, nobles, and great ladies of the time. King James VI of Scotland was a guest. Also often featured was one of Tycho's favorite pets, a fully grown elk. Occasionally, Tycho would order silence so that the musings of a dwarf named Jepp could be heard. Jepp lived under the banquet-hall table.

Tycho was a hard-headed, hard-drinking aristocrat. One Christmas evening, his hot temper and argumentative disposition got him into a drunken duel with another Danish nobleman who was a little less drunk and who cut off Tycho's nose. From that time forward, Tycho wore a variety of copper and gold nose prosthetics which he painted flesh color.

Despite his wild lifestyle, Tycho had a mission. He owned copies of every astronomical table that was available, and what he discovered was that every one of them was faulty. "I've studied all available charts of the planets and stars," he wrote, "and none of them matches the others. There are just as many measurements and methods as there are astronomers and all of them disagree. What's needed is a long-term project with the aim of mapping the heavens conducted from a single location over a period of several years."

So he dedicated the part of his life not devoted to drinking to building the best astronomical database in the world. Tycho, without the aid of a telescope, eventually charted more than 777 stars. Not satisfied with either the Copernican or Ptolemaic model, he developed his own scheme for how the Universe was arranged.

MARTIAN KITSCH

On any given day, more than 3,400 Mars-related items are offered for sale on eBay. They range from books, toys, collectibles, soundtracks, DVDs, T-shirts, and pinball machines to a pair of Michael Jordan "Retro Mars" Nike shoes available for $700.00.

An Internet search using Google pulled up 347 million references to Mars, including everything from a huge seaplane called the *Martin Mars* to chocolate Mars Bars. There's a Mars Hall, a Field of Mars, even a Mars Society. And then, of course, there are all those science web sites dedicated to the Fourth Planet. Where did this dizzying array of stuff come from?

A novelist is to blame.

In the H.G. Wells story, *The War of the Worlds*, Martian invaders launch an attack against Earth only to get cut down by terrestrial microbes.

An ambitious young director, Orson Welles, and his Mercury Theater group produced an adaptation of the story told as if it were a series of live news bulletins. When it was broadcast on the night of Halloween 1938, many people believed the bulletins were real. Whole families wrapped their faces in wet towels to protect themselves from Martian gas. A group of scientists from Princeton set out to find the reported crash site. Anxious churchgoers gathered to pray in Harlem. *The New York Times* received more than 800 calls requesting information or to relay the report they had just heard. It has been estimated that over 1.2 million people were thrown into panic by the show.

"Ladies and gentlemen, I have a grave announcement to make. Incredible as it may seem, strange beings who landed in New Jersey tonight are the vanguard of an invading army from Mars."

— RADIO BROADCAST OF THE H.G. WELLS CLASSIC, *WAR OF THE WORLDS*, OCTOBER 31, 1938

The infamous radio broadcast established Orson Welles as an *enfant terrible* of theater and film.

Welles's hoax sparked a mania for all things Martian. Edgar Rice Burroughs of Tarzan fame tapped into the surprising new market with his John Carter "Man of Mars" series of books — a favorite of Carl Sagan when he was a boy.

Hollywood found gold on Mars, especially during the Cold War. The George Pal version of *The War of the Worlds* is a film classic. It was followed, in true Hollywood fashion, by sequels: *Invaders from Mars, Mars Attacks, Robinson Crusoe on Mars, Angry Red Planet* are all science-fiction favorites. Marvin the Martian always got the best of Daffy Duck. Even Abbott and Costello went to Mars.

The space race also contributed to Mars mania. Iconic V-2-type rockets dotted the landscape of films like *When Worlds Collide* and episodes of the television show *The Outer Limits*. Television had its own answer for the Mars craze, with *My Favorite Martian, Lost in Space,* and *Star Trek*.

Even the world of music has something to say about Mars, from Gustav Holst's *The Planets Suite* to David Bowie's "Spiders from Mars" and Elton John's "Rocket Man."

In the 1990s a new round of Mars movies, just a tad more sophisticated than their predecessors, appeared. Val Kilmer's *Red Planet* tells the story of a group of astronauts stranded on Mars. Brian De Palma's *Mission to Mars* plays with the urban legend surrounding the Face on Mars.

From the "Sailor Moon" cartoon series to "Doom" video games, Mars has become our vision of thrills and perils to come.

"No one would have believed in the early years of the twenty-first century that our world was being watched by intelligences greater than our own...." Mars invaded Earth's cinemas again in 2005 with Steven Spielberg's remake of the celebrated *War of the Worlds*. The story is as much about the fragility of human civilization as it is about invaders from Mars. An earlier (1953) version by George Pal exposed the technological weakness of our civilization when the atomic bomb, delivered by a futuristic aircraft, failed to stop the Martian tripods. Spielberg's version is all the more disturbing because we see the collapse of civility, when a mob swarms the car driven by Tom Cruise's character. This illustration is a modern version of the Martian tripods originally created by Alvin Correa for the 1906 French edition of the book, *Guerre des Mondes*.

In order to validate his ideas, Tycho needed to match his observations with the motions of the planets, which would allow him to make predictions about their future locations. So he set about building the most complete database for the planet with the greatest *parallax* — Mars. (Parallax is the optical phenomenon that causes things that are nearby to appear to move faster as you pass them than things that are far away. For example, fence posts along a road seem to whip by, while a telephone pole in the distance appears to be stationary, even though you're moving at a constant speed. Mars has the greatest parallax because it is like the fence posts in relation to other planets, such as Jupiter and Saturn, which are farther away.) Tycho charted Mars's oppositions, motions, and parallax, all from naked-eye observations, for he was the last and perhaps the greatest of the pre-telescope astronomers. His observations were accurate to within 1 arc minute, which is one-sixtieth of a degree. Tycho was able to build an exquisite data set, but he lacked the theoretical skills to make the data match his model. He needed help, and he knew it.

Johannes Kepler could not have been more unlike Tycho had he tried. Pious, humble, and poor, Kepler wanted initially to be a theologian. But like Tycho, Kepler became interested in astronomy at an early age, and as a young man sought to prove his own theories about the geometry of space. If only he had the data!

In 1596, Kepler published his first work, *Mysterium Cosmographicum*, in which he proposed that the planets were arranged according to the nested geometry of five perfect solids. These were the Pythagorean solids: three-dimensional forms which consisted of a sphere, a three-sided pyramid, a cube, an octahedron, a dodecahedron, and an icosahedron. The model was a failure. It was scorned by Galileo, the Copernicans, and the traditionalists alike, but the brilliance of its author was recognized in Uraniborg.

By 1599, Tycho and his entourage had relocated to Prague, and soon Kepler received an invitation. Kepler had already been the victim of religious persecution in Germany, and he immediately accepted Tycho's offer. In 1600, he became Tycho's assistant.

Things did not go smoothly. Kepler hated the raucous parties, and Tycho worried that the quiet, pious Kepler would run away with his data. Tycho dribbled just enough information to keep Kepler working on the problem of Mars but never enough to permit a breakthrough. Their seismic relationship lasted only a year. Tycho died suddenly from his excesses at the table. His last words were, "Let me not seem to have lived in vain."

There was some speculation that Kepler had murdered Tycho, but these allegations have never been proved. He did, however, acquire the priceless data. Finally he was in possession of the information he needed and went about the analysis of the motions of Mars with renewed vigor.

His first task was to discover the shape of Earth's orbit, which he did by triangulating Earth's position between Mars and the Sun. Einstein hailed Kepler's use of Mars for this triangulation as a stroke of genius. What Kepler discovered was that Earth's orbit was, as the Copernican model suggested, nearly circular (although the center of the circle, it turns out, is offset from the center of the Sun).

Now Kepler found that any predictive theory based on circular orbits had the position of Mars deviating from Tycho's observations by as much as 8 arc minutes. He knew Tycho's data was accurate to within 1 arc minute, so something had to be wrong. Kepler also discovered that Mars did not move across the sky at a constant rate: Instead, the planet appeared to speed up and slow down — an immediate contradiction to the Copernican assumption that planets moved at a steady pace. These changes in speed could not be explained away by Ptolemaic epicycles.

In 1609, Kepler published his findings in *Astronomia Nova*, in which he trumpeted the discovery of his first two laws of planetary motion. The first law states simply that the shape of a planet's orbit is an ellipse, with the Sun at one of the two focal points inside the ellipse. The second law states that planets moving along their orbits sweep out equal areas in equal times.

The riddle of planetary motions was solved, but Kepler was sorely disappointed by his findings. He had wanted so much to find a perfect geometry for the cosmos. Instead of symmetry, perfection, and harmony, he found a messy collection of whirling ellipses. He expressed his bitterness in a letter to a Uraniborg alumnus, Tycho's assistant Longomontanus. "I have cleared the stable of astronomy of cycles and spirals," he wrote, "and all I have left is a cartful of dung."

CHARTING THE LANDSCAPE OF MARS

In the same year that *Astronomia Nova* was published, the Italian astronomer Galileo Galilei copied a design for a new optical device that had been developed in Holland the previous year by an eyeglass manufacturer, Jan Lippershey. Lippershey's device was a pair of *occhiali* — refracting glass lenses — aligned with each other at the opposite ends of a tube. This device allowed very distant objects to be seen as if they were up close. The design was not complicated, and legend has it that Galileo copied it in a single evening.

Galileo turned his new telescope toward the Moon and immediately discerned hills and valleys just like those on Earth. And then, while at the University of Padua in 1609, he became the first human to make telescopic observations of Mars. Others soon were making similar observations of their own.

In 1659, the Dutch astronomer Christiaan Huygens made a sketch of Mars featuring the region now known as Syrtis Major. Seven years later, the Italian astronomer Giovanni Cassini estimated to within 3 minutes the length of a Martian day, which he reported as being equivalent to 24 Earth-hours plus 40 Earth-minutes. Cassini is also credited with the discovery of the north pole of Mars.

Perhaps taking a cue from Copernicus, Huygens arranged for an explosive manuscript he wrote in 1695 to be published 3 years later — after his death. *Cosmotheoros* contains Huygens's daring ruminations about the nature of life on other planets in our solar system. He called the denizens of other planets *planetarians* and speculated that at some level they must behave and organize themselves in a manner familiar to humans.

Huygens was by no means the last astronomer to speculate on the possibility of life on the Red Planet.

THE GALILEO CODE

Nearly three decades later, the first of a series of Martian enigmas emerged. When Jonathan Swift published his satirical masterpiece *Gulliver's Travels* in 1726, he correctly described the two small moons of Mars and gave a fairly accurate approximation of their relative distances from the Martian surface.

"They [the Laputians] spend the greatest part of their lives in observing the celestial bodies," he wrote. "They have likewise discovered two lesser stars, or satellites, which revolve about Mars." What makes this intriguing is that the moons of Mars weren't discovered until roughly 150 years later. There was no way that Swift could have known about them, and yet he was correct. Was it a lucky guess?

The real face of Mars is scarred, pocked, cratered, and crevassed. Visible on the left are the dark Tharsis Montes volcanoes (Pavonis Mons at the bottom left, Ascraeus Mons above). In the center is the massive canyon complex of Valles Marineris (near bottom of image) and on the right, the rugged hills of Xanthe Terra.

One explanation may involve an intrigue worthy of a sequel to *The Da Vinci Code*. Let's call it *The Galileo Code*.

It seems that back in 1610, while experimenting with his new telescope, Galileo made a discovery that he was unsure would bear scrutiny. In order to hedge his bet, so to speak, he needed a placeholder, something that would register the discovery yet withhold it until there was positive verification. So, he created an anagram (a word or phrase in which the letters are transposed) and distributed it to notables all over Europe. In this case, the announcement read as follows:

s m a i s m r m i l m e p o e t a l e u m i b u n e n u g t t a u i r a s

Included on the distribution list was the upstart Johannes Kepler. Now, there was no way that Kepler was going to let a puzzle like this languish on the shelf. He became convinced that the solution to Galileo's riddle was "*Salue umbistineum geminatum Martia proles*" (Hail, twin companionship, children of Mars).

If Kepler was right, then Galileo was the first to observe the moons of Mars. We'll never know for certain, but Galileo would have needed a far more powerful telescope than the one he had to observe the tiny Martian moons, so it seems likely that Kepler was wrong. The correct answer, discovered some decades later, is almost certainly "*Altissimum planetam tergeminum observavi*" (I have observed the most distant planet to have a triple form). The "distant planet" is Saturn.

Still, the notion that Mars had two moons gained favor in scientific circles. In fact, the two moons of Mars became something of a staple in eighteenth-century science fiction. The great writer and philosopher Voltaire, in his 1752 short story "Micromegas," gives us a glimpse into the logic of a two-moon Mars: "Those excel-

lent philosophers know how difficult it would be for Mars, which is so distant from the Sun, to get by with less than two moons."

If Venus and Mercury had no moons, Earth had one moon, and Jupiter had, as was known at the time, four Medicean moons, then a numerical progression of doubling would suggest that Mars has two. This notion was obviously a relic of the quest for harmonic or geometric perfection, but in the odd case of Mars, it actually yielded the right answer. As the American astronomer Asaph Hall was to discover more than 100 years later, Kepler, Swift, and Voltaire had all gotten it right!

A CANTICLE FOR THE MARTIANS

In 1781, Sir William Herschel, a noted composer of classical music who had no technical background beyond an amateur's love of mathematics, discovered a new planet and named it Georgium Sidus (George's Star) in honor of King George III. France was not about to abide a cosmic landmark named after an English king, so Uranus, chosen in honor of the same muse that had inspired the construction of Uraniborg nearly 200 years earlier, became the accepted name. King George made Herschel the royal astronomer.

That same year, Herschel discovered that Mars is tilted 25 degrees, similar to Earth's tilt of 23 degrees. This meant that Mars, like Earth, must have seasons. In fact, while similar, the more nearly circular orbit of Earth tends to moderate our seasonal changes, whereas the elliptical orbit of Mars makes them more extreme.

Herschel tried to describe accurately the nature of Mars's polar caps. These had already been observed by Cassini's nephew, Giacomo Maraldi, in 1719. Maraldi spotted a dark ring that runs along the perimeter of the *taches blanches* (white spots), and suspected that it was caused by running water from a melting ice cap. Herschel's observation reinforced these suspicions.

By 1784, Herschel had explored the issue of whether or not Mars has any form of atmosphere. He tracked two stars as they disappeared behind the planet. If there was a thick atmosphere, the stars should dim slowly as they passed behind the disk of the planet. If the atmosphere was thin, only the slightest dimming would occur at the last moment. If there was no atmosphere — a conclusion that the existence of polar caps would contradict — then the stars would not dim down at all but rather wink out. What Herschel observed was in fact a last-minute, slight dimming of the stars, thus proving the existence of a thin atmosphere.

Eighteenth-century astronomers had established a number of fundamental truths about Mars. They calculated its orbital period and the length of its day. They observed its axial tilt and the existence of its seasons, its polar cap and even its atmosphere. The distance of Mars from the Sun and Earth now could be calculated. But disputes over the topography continued.

THE AREOGRAPHERS

With the principal features of Mars now established, the challenge was to chart the planet's surface. This was difficult because telescope optics were so poor and Mars is so small and far away. Mars appeared to be constantly changing. At times, this may actually have been the case: Global dust storms do occasionally obscure the Martian landscape. But, by 1830, the improvements in telescope optics needed for serious, systematic investigation had become available.

Work began in earnest when Johann Madler and Wilhelm Beer, a pair of astronomers/cartographers famous for their work in charting the Moon, turned their attention to the Red Planet. They watched the southern polar cap on Mars expand and recede with the changing seasons. By 1832 they designated the first prime meridian for Mars.

The *prime meridian* is the first of the grid lines that run from pole to pole on a globe. These lines are called *longitude* lines, and they are used to establish the east-west coordinates of a place or thing. They are perpendicular to lines of *latitude*, which run parallel to the equator. It is obvious that the lines of latitude should start at the equator and end at the pole, but longitude lines are set arbitrarily. Curiously, Mars's prime meridian was assigned about 50 years before Earth's, which wasn't established until 1884.

The patriarch of a family dynasty of astronomers, William Herschel (1738–1822) discovered the planet Uranus and two of its moons, as well as two moons of Saturn.

During the planetary opposition (that is, its closest approach to Earth) of 1858, a Jesuit priest, Father Angelo Secchi, made a new series of sketches of Mars. Secchi named all the features he could identify after famous astronomers, but as fate would have it, he applied to one of its features the suggestive term Atlantic Canale. He either guessed or assumed that this dark region was a body of water separating an old world from a new world, just as on Earth. Later, the Atlantic Canale region would be known as Syrtis Major. He identified other features on Mars, including continents, oceans, peninsulas, and bays galore, from Cassini Land and Copernic Continens to Newton Ocean and Beer Sea. And of course there was a modest Secchi Continens occupying the area at about 30 degrees latitude and 45 degrees longitude near the north pole. By 1869, Secchi had described other features on his map as *canali* as well.

The science of areography, which is the study of Martian landforms, just as geology is the study of landforms on Earth, had begun in earnest. The word *geology* is derived from the Greek *geo*, which is in turn related to *gaia*, meaning "the living Earth," and *areography* is derived from the name of the Greek god of war, Ares.

THE CANALS OF MARS

Giovanni Schiaparelli was an astronomer at the Brera Astronomical Observatory in Milan for 40 years. For 38 of those years, until he retired in 1900, he was the observatory's director. Schiaparelli used his time at the observatory to draw a map of Mars that far exceeded in accuracy everything that had come before it. But it was the nomenclature Schiaparelli chose from classical antiquity — place-names like Elysium, Tyrrhena, Hesperia, and Cimmerium — that imbued his map with a sense of mystery.

These are places we feel we have heard about and want to see.

But what really titillated the public's imagination were those long, stringy *canali*.

It is no secret that Schiaparelli's use of the term *canali* was not intended to describe artificial constructs on the Martian surface. He meant simply to label what he perceived to be channels, canyons, and gullies. Yet, in the era of massive canal building here on Earth — the Suez and Panama canals were either being planned or already under construction in Schiaparelli's lifetime — it is easy to understand how the word was so famously mistranslated.

"There are on this planet, traversing the continents, long dark lines which may be designated as canali, although we do not know what they are," wrote Schiaparelli in 1882. "Their arrangement appears to be invariable and permanent; at least as far as I can judge from four-and-a-half years of observation."

Despite his vague denials, a powerful mythology emerged from his handiwork. By 1879, the book publishing industry had spotted a lucrative franchise. That year, a Professor Schmick published a book in Cologne entitled *Der Planet Mars, eine zweite Erde (The Planet Mars: A Second Earth)*. A monograph entitled *Mars, eine Welt im Kampf ums Dasein (Mars, a World Engaged in the Struggle for Survival)*, by Otto Dross who, doubtlessly influenced by Darwin, described the toil of Martian engineers in a race to reclaim arable land from the encroaching and relentless deserts. Both titles were classified as nonfiction. In 1898, that pillar of science fiction and Martian lore known as *The War of the Worlds* was published. This H. G. Wells masterpiece has inspired books, movies, magazines, rock opera, theater, radio broadcasts — you name it. The primal appeal of *The War of the Worlds* had lodged itself permanently in our collective psyche. And it's all the result of those canals!

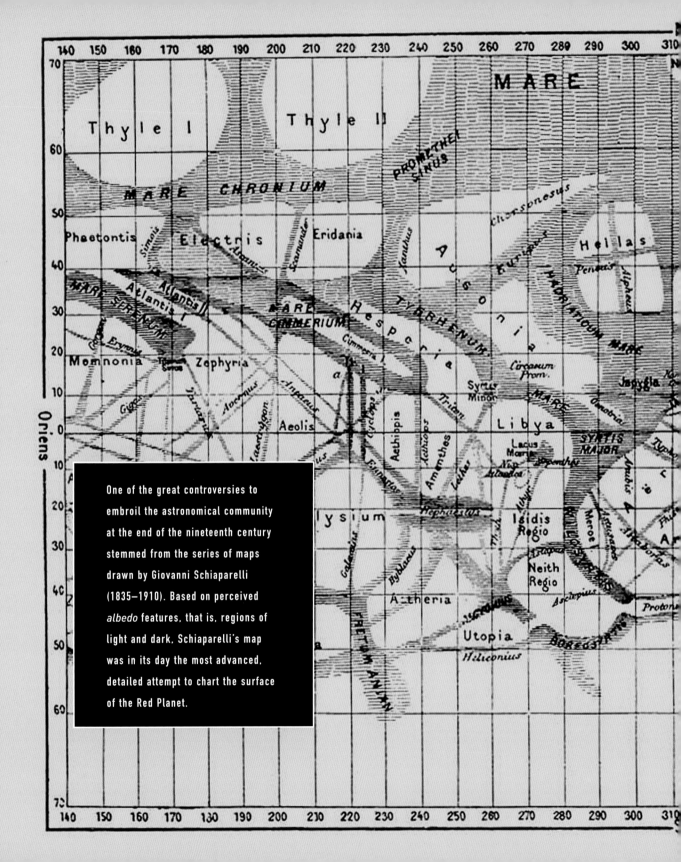

One of the great controversies to embroil the astronomical community at the end of the nineteenth century stemmed from the series of maps drawn by Giovanni Schiaparelli (1835–1910). Based on perceived *albedo* features, that is, regions of light and dark, Schiaparelli's map was in its day the most advanced, detailed attempt to chart the surface of the Red Planet.

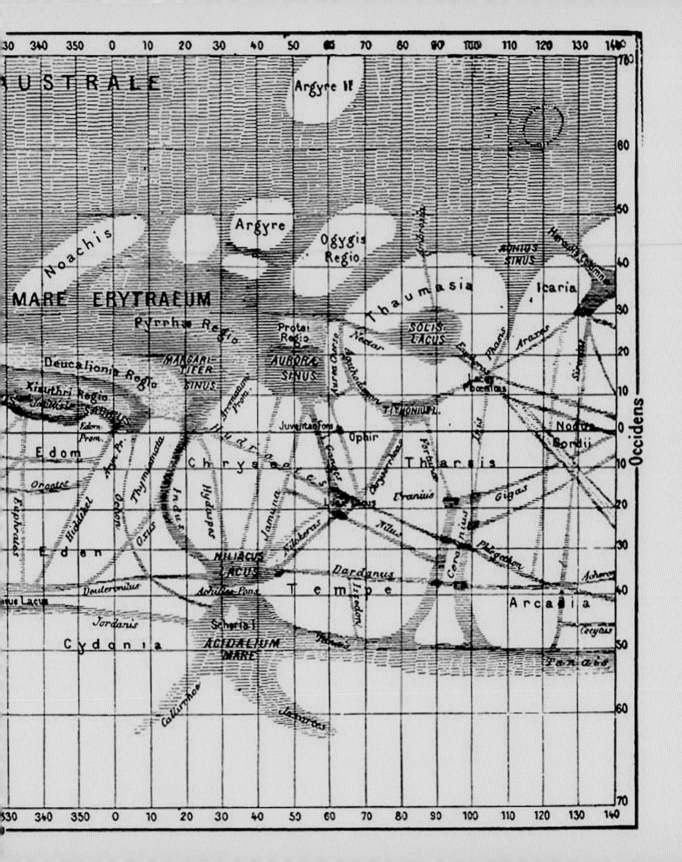

MEET PYGMALION

Percival Lowell is a name that has become synonymous with Mars. The son of a Massachusetts industrialist, Lowell graduated from Harvard in 1876 but turned down a teaching position in mathematics there. He chose instead to travel in the Far East, in search of himself and the exotic. Much of his time overseas was devoted to the study of Japanese culture. His scrutiny of the Japanese reinforced his belief in social Darwinism, that is, the notion that societies and nations are like living organisms in a struggle to survive.

By the time he was 40, Lowell was bored with the Far East and thirsted for something else. That thirst was quenched when he obtained a copy of Camille Flammarion's *La Planete Mars*, published in 1892. Flammarion argued for the existence of Schiaparelli's *canali* but pushed further and defended them as artificial constructs, similar to Roman aqueducts, and built by an older, superior race of beings.

Lowell beheld his life's work. He had seen the maps and heard the debates. For Lowell, the *canali* were there and the plight of those canal builders was real. So, with the help of the astronomer William H. Pickering, Lowell set up an observatory in Flagstaff, Arizona, atop a 7,500-foot (2,300-meter) mesa Lowell dubbed Mars Hill. Lowell Observatory began operations in 1894, in time for the Mars opposition later that year.

At the same time, Lowell began writing the first of three books — *Mars* (1895), *Mars and Its Canals* (1906), and *Mars As the Abode of Life* (1908) — in which he marshaled and refined the evidence for extraterrestrial life on Mars. The *canali* visible from Earth were no longer the actual canals but rather the vegetation growing along the banks of the canals. At the places where the canals intersected, larger spots could be seen where the vegetation

"Startling as the outcome of these observations may appear at first, in truth there is nothing startling about it whatever. Such a possibility had been quite in the cards ever since the existence of Mars itself was recognized by the Chaldean shepherds."

— PERCIVAL LOWELL, *MARS*, 1895

was evidently spread farther out. Pickering was the first to report the spots, and Lowell believed he saw them as well.

Scientists often speculate about the meaning of their data; in fact, that is one way by which a science is advanced. But Lowell's imagination got the best of him. To Lowell, as he looked at these crisscrossing shadows, it appeared that the Martians were locked in a battle against the encroaching desert in a world gone dry.

The Martian north pole is covered by water ice, unlike the south polar ice cap which consists mostly of frozen carbon dioxide (better known as *dry ice*). This picture of north polar ice covers an area approximately 1.5 miles (1 kilometer) wide, and was taken September 7, 2004, by *Mars Global Surveyor*.

Percival Lowell (1855–1916) was utterly convinced that Mars was inhabited by an advanced alien civilization. The sheer determination and vigor with which he pursued this notion framed late nineteenth- and early twentieth-century debate on the subject.

Lowell's books were popular. Indeed, he became something of a celebrity, and his appearance filled lecture halls everywhere. The plight of the Martians captured the world's imagination, and people flocked to hear him speak.

But Lowell's Martians had detractors, too. Leading the anti-Martian charge was Alfred Russell Wallace, coauthor of the theory of evolution. In 1907, Wallace published *Is Mars Habitable?* which started as a review of the second of Lowell's books, *Mars and Its Canals*. It grew into an unequivocal rejection of the canals, the Martians, and Lowell. Wallace cited the thin atmosphere, the lack of water, and the extreme cold as factors that collectively ruled out any chance for indigenous life. He went even further, insisting that life could not exist beyond Earth or in any place else in the Universe, a position shockingly at odds with the theory Wallace helped develop. But he was not alone. Even Schiaparelli was tepid in his response to Lowell's claims.

To his dying day, Percival Lowell believed in his Martians and their struggle, and much of the rest of the world did too. His speculation had ranged far beyond the available evidence. He was no longer merely an observer; he had become the inventor of a life form. The Martians were the children of Pygmalion, and they had become an example for Earth to follow.

The Martian craze of the nineteenth and twentieth centuries coincided with other dreams, for example, of artificial waterways that connected oceans, and of railroads that joined the edges of continents. It was an age when new discoveries were made almost daily. Lowell's Martians inspired decades of pop literature and films and motivated our quest to send reconnaissance spacecraft into Martian orbit and onto Martian soil. Technology has, for the moment, exiled the Martians to the gulag of lost causes. But in the end, the Martians may have the last word.

MARS IN THE SPACE AGE

"AND EARLY IN THE TWENTIETH CENTURY CAME THE GREAT DISILLUSIONMENT."
— H. G. Wells, *The War of the Worlds*

Our fascination with Mars helped form the background for dreams of space travel. Inventors and cranks put together the first small rockets that would one day become the giants that made access to space possible. But the dream of finding life on the Red Planet was only one inspiration. The other was a nightmare vision of rockets as weapons launched against terrestrial enemies. Surely Mars, the Greek god of war, would have been pleased.

NASA's *Opportunity* rover captured this forbidding image of Victoria Crater in 2006.

THE AMERICAN PIONEER

Few would have guessed by looking at him that Robert H. Goddard would become a towering figure in the history of rocket science.

Home-movie footage reveals him as an amiable, balding man. He is seen tinkering with one of his trapezoidal contraptions in a scene reminiscent of one showing the early twentieth-century eccentrics lampooned in the film *Those Magnificent Men in Their Flying Machines*. His original design was based on two patents he registered in 1914, one month before the outbreak of World War I. His earliest experiments were conducted on his Aunt Effie's farm in Auburn, Massachusetts. A liquid-fuel rocket launched in 1926 attained a height of about 41 feet (12.5 meters) in the air and traveled 182 feet (55.5 meters) downrange — an inauspicious flight to be sure, but it demonstrated the efficacy of the fuel — and he pressed on. A later test in 1929 was so loud that it brought emergency vehicles to the farm. The headline in the local paper the next day made fun of the event: "Moon Rocket Misses Target by 238,799 $\frac{1}{2}$ Miles." Goddard was barred by local authorities from conducting further test flights in the neighborhood, and so he moved them to a more remote location at Hells Pond, near Camp Devens, Massachusetts.

As a boy, Goddard was fascinated by fireworks, enthralled by the Martians imagined by H. G. Wells, and dazzled by the vision of Martian life promoted by Percival Lowell. He dreamed of building a rocket that could take humans to the Moon and on to Mars. Just as the Wright brothers had single-handedly addressed the problem of heavier-than-air powered flight, Goddard tackled every basic problem involved with rocketry, from self-cooling engines to the lack of air in space (which he remedied by supplying his rocket with liquid oxygen).

"As I looked toward the fields in the east I imagined how wonderful it would be to make some device which had even the possibility of ascending to Mars."

— ROBERT H. GODDARD, DIARY ENTRY, QUOTED BY EDNA YOST

A voracious reader, Robert H. Goddard (1882–1945) dreamed of Mars after reading H.G. Wells' classic *The War of the Worlds*. He is seen here in his workshop tinkering with one of his rockets and perhaps dreaming of interplanetary journeys.

In 1929, Goddard set out to raise funds for more extensive research. He chose a difficult time — the Great Depression was just beginning. But, with the help of aviation pioneer Charles Lindbergh and Lindbergh's friend, the industrialist and philanthropist Daniel Guggenheim, Goddard secured the funding he needed to move to Roswell, New Mexico. He would pursue his dream in the desert.

During World War II, Goddard's lab was penetrated by a spy from the German intelligence agency Abwehr. The spy sent back to Germany many designs for parts that Goddard later recognized when he inspected German rockets brought stateside. How he would have loved to have compared notes with his Axis counterparts after the war! But this was not meant to be: Robert Goddard was diagnosed with throat cancer at Johns Hopkins Hospital in Baltimore and passed away shortly thereafter in 1945. In 1959, the brand-new space agency, NASA, dedicated the Goddard Space Flight Center just outside Washington, D.C., to the father of rocketry.

FROM ROCKET CLUBS TO WAR

In Europe, the early development of rockets took place in rocket clubs. None was more advanced than the Rocket Club in Germany, the Verein für Raumschiffarht. After members of the club demonstrated a liquid-oxygen, gasoline-fed rocket motor in 1930, the club secured funding from the German government. Their work soon surpassed Goddard's.

The first rocket launched from Cape Canaveral, on July 24, 1950, was a hybrid that combined a German V-2 missile with an American Corporal rocket. This Florida test was called *Bumper 8*. It was a rarity: the Americans conducted most of their post-war tests of the V-2 at White Sands, New Mexico.

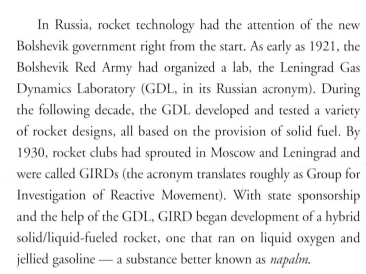

In Russia, rocket technology had the attention of the new Bolshevik government right from the start. As early as 1921, the Bolshevik Red Army had organized a lab, the Leningrad Gas Dynamics Laboratory (GDL, in its Russian acronym). During the following decade, the GDL developed and tested a variety of rocket designs, all based on the provision of solid fuel. By 1930, rocket clubs had sprouted in Moscow and Leningrad and were called GIRDs (the acronym translates roughly as Group for Investigation of Reactive Movement). With state sponsorship and the help of the GDL, GIRD began development of a hybrid solid/liquid-fueled rocket, one that ran on liquid oxygen and jellied gasoline — a substance better known as *napalm*.

With the onset of World War II, the Germans, Russians, and to a lesser extent the Americans were each well positioned to study the military possibilities of rocketry. All three developed small, unguided rockets that were used, in effect, as artillery. The Germans deployed their Nebelwerfer rocket launcher for the first time against the Russians in the summer of 1943 during a massive tank battle at Kursk. The Russians replied with their infamous Katyusha, which was not very accurate but in saturation bombing was effective against the Germans. The Katyusha was so successful, it is still in use in the Middle East today. None of the other combatants in World War II, however, came close to matching the technical achievements of the Germans.

THE V-2

The Vergeltungswaffe 2 (V-2) was the world's first operational ballistic missile. Tapering to a point at the top, slightly bulging in the middle, with *Frau im Mond* fins at the bottom, the V-2 remains unchallenged as the iconic image of rocket science.

On January 24, 1945, the Germans successfully tested the A-9, the world's first missile designed to have intercontinental reach. It flew to an altitude of 50 miles (80 kilometers) and could easily have carried out the purpose for which it was designed — to bomb New York, Chicago, or Washington. There were even plans for a manned three-stage rocket, designated the A-11. The A-9 never flew a second time, and the A-11 never got off the drawing board, but more than 20,000 V-1 and V-2 vengeance weapons had been launched by the time the Third Reich collapsed.

One of the principal centers where the V-2 was manufactured was an underground factory at Mittelbau-Dora, near Nordhausen, a small city in central Germany. When Mittelbau-Dora was liberated in April 1945, a dark truth about the V-2 was uncovered: It had been built by slave laborers primarily from Bergen-Belsen but also from Auschwitz, Buchenwald, and other Nazi death camps. Harsh treatment and dreadful working conditions caused more people to die from building the V-2 than from being attacked by it. The dreams of German rocket scientists were built on the nightmare realities of the Holocaust.

At the close of World War II, there was a mad scramble for leftover V-2 rockets, parts, manufacturing tools and dies, and for the scientific expertise that had designed and built them. Many German scientists sought and surrendered to the Americans rather than risk capture by the Soviet Red Army, which they feared would mistreat them. Prominent among the new émigrés to America was the scientist Wernher von Braun. His probable awareness of SS activities at Mittelbau-Dora have cast a dark shadow on his reputation, but his capture by the Americans and his relocation to the United States presented a fresh start for American rocket research.

Sputnik I shocked the world when it was launched on October 4, 1957. Most of it burned up on re-entry but a few pieces landed in the yard of a private residence near Los Angeles.

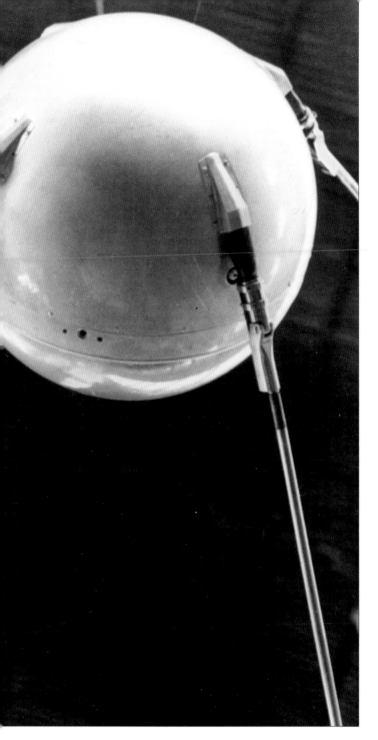

Von Braun moved to Fort Bliss, Texas, just outside El Paso, where he helped in the reconstruction and test firing of captured V-2 rockets at the White Sands Missile Range in New Mexico. From 1946 until 1951, some 68 modified and improved V-2 rockets found their way into the skies over the American desert. Some reached altitudes of 128 miles (206 kilometers) above Earth, more than double their wartime zenith. In one experiment, researchers were able to measure the ionization that occurs beyond Earth's protective atmosphere and, in another, with all noise-making systems turned off, scientists could listen to the sounds of micro-meteor dust grains striking the outer surface of the rocket. For the first time, humans were testing the water on the shoreline of space.

HELMUT GROETTRUP

The Russians also benefited from the expertise of German scientists. In the final months of the war, the Red Army gathered up roughly 1,000 V-2 rockets from a number of rocket test facilities in eastern Poland along with scores of scientists and technicians.

One team of 72 scientists, headed up by von Braun's former assistant, Helmut Groettrup, was moved to Gorodomlya Island in the middle of Seliger Lake in northwestern Russia. Much of what we know about the work on Gorodomlya comes from the diary of Groettrup's wife, Irmgard. She described the conditions on the island as austere, with neither running water nor sanitation when they arrived, but things improved later. Although they were paid, they were essentially imprisoned on the island and not permitted to leave until their repatriation to East Germany in November 1953.

In 1947, Groettrup and his wife attended the launch of 11 V-2 rockets at the secret Soviet firing range at Kapustin Yar, near Stalingrad (now known as Volgograd). By the following year, Groettrup's team had produced an improved version of the V-2 designated the G-1.

Unknown to Groettrup, his team on Gorodomlya was in direct competition with a Soviet team, headed by Sergei Korolev, to develop an intercontinental ballistic missile (ICBM). The design specifications required that the rocket be capable of delivering a 6,600-pound (3,000-kilogram) payload — meaning an atomic bomb — a minimum of 1,800 miles (3,000 kilometers) downrange. This would place every target in western Europe within reach. Groettrup introduced his multistage G-4 design, which compared favorably with Korolev's competing R-3 design; in fact, the G-4 was better. But, in the end, only a wind-tunnel model of the G-4 was ever built. Its conical shape, however, was an innovation echoed in Korolev's subsequent design for the R-7, which became the world's first operational ICBM. The R-7 has become a workhorse that still carries cosmonauts to the *International Space Station* today.

"It seemed as though all the gates of Hell opened up...the vehicle agonizingly hesitated for a moment, quivered again and in front of our unbelieving, shocked eyes, began to topple. It sank like a great flaming sword into its scabbard... hitting the ground with a tremendous roar that could be felt and heard even behind the two-foot concrete walls of the blockhouse."

— KURT STEHLING, *VANGUARD* ENGINEER

The *Vanguard* rocket was launched in haste on December 6, 1957. Roughly 2 seconds into the flight, it lost power, dropped back onto the launch pad, and its fuel tanks exploded.

FELLOW TRAVELER

The Soviets called the first artificial satellite launched in space *Sputnik* (meaning "fellow traveler"). What they had in mind was a companion to Earth in its journey around the Sun. But they also must have known that for Americans who had lived through the McCarthy era, a fellow traveler was a Communist sympathizer.

The 184-pound (83.5-kilogram) *Sputnik I* was launched on October 4, 1957, and flew around Earth for 4 months before its orbit decayed and the satellite burned up in Earth's atmosphere. But even before the first *Sputnik* faded away, the Soviet Union launched a second, on November 3. *Sputnik II* weighed in at more than 7,000 pounds (3,100 kilograms), but it wasn't just its size that impressed the West: *Sputnik II* carried a passenger, Laika, the first dog launched into space. The Communists had beaten America into space. Twice.

In a national address broadcast on November 7, 1957, President Dwight Eisenhower tried to reassure Americans that the United States was not all that far behind the Russians. To underscore his point he displayed a prop, the nose cone of a Jupiter C missile (a modified *Redstone* rocket). This was followed a month later by the live, televised launch of America's first satellite aboard a *Vanguard* rocket. The United States wanted to make a point about the openness of its space program in contrast to the secrecy of the Soviet effort. And so, on live television, before millions of viewers, the slender *Vanguard*, instead of lifting off, slid backward into a catastrophic ball of fire.

DAS MARSPROJEKT

Around 1948, as the V-2 program was winding down, Wernher von Braun wrote a novel about an expedition to Mars. The story was supplemented by a set of appendices that formed the engineering basis of a technically precise, comprehensive study for a mission to Mars. What von Braun set forth was not some modest, low-budget reconnaissance flight. He envisioned a massive expedition, with a flotilla of ten vehicles, each weighing 4,100 tons (3,720 metric tons). This was serious stuff.

Each vehicle would be built in low Earth orbit by means of a space shuttle operating from a base on Johnston Island, a coral atoll in the Pacific that would later become the site of nuclear weapons testing by the United States military. Von Braun's shuttles consisted of three stages which would all be recovered and reused. The first-stage booster would parachute downrange 189 miles (304 kilometers); the second stage would parachute 907 miles (1,459 kilometers) downrange; the third stage, which was the shuttle itself, would rendezvous with the flotilla and deliver 28 tons (25 metric tons) of cargo and 16 tons (14.5 metric tons) of nitric acid/hydrazine fuel. Von Braun estimated that there would have to be 46 shuttles in the shuttle fleet, and that 950 shuttle flights would be needed.

Once the flotilla was completed, 70 astronauts would board the ships. The ships would ignite their main engines, consuming 76 percent of the fleet's fuel during an initial burn. After 66 minutes, the engines would shut down and the fleet would coast for the next 260 days. When the fleet arrived at Mars, they would fire their engines again and enter a circularized orbit. They would then search for the best site for a Mars base, to be established somewhere along the Martian equator.

Von Braun imagined that a large glider would descend from the fleet and land on skis at the north pole of Mars. This glider would then deploy several crawlers that would trek from the Martian ice cap toward the designated position along the Martian equator — a 4,000-mile (6,500-kilometer) journey.

Once the crawlers reached their destination, a base would be set up and a landing strip would be cleared, allowing additional crew to land in two more gliders. A total of 50 personnel would be stationed on the ground, while the remaining 20 would remain in orbit with the flotilla.

On the ground, the astronauts would survey "the flora and fauna" of Mars. It was still believed, when von Braun was writing, that some sort of life form might exist on Mars, perhaps even a dying race of canal builders. Lowell's vision was still burning brightly well into the 1950s.

No one was to be left behind in von Braun's novel. His scheme made provision for the crew's safe return to Johnston Island.

Although Chesley Bonestell (1888–1986) made his living as a matte painter for Hollywood motion pictures, his iconic visualizations of the exploration of space were what fired the public's imagination.

America's humiliation was complete.

President Eisenhower turned to von Braun's team, which had been quietly working in Redstone, Alabama. On January 31, 1958, a U.S. Army Jupiter C rocket thundered to life and lifted the 30-pound (14-kilogram) *Explorer I* satellite into space. The satellite contained a science package developed at the University of Iowa by James Van Allen, and it made the first new discovery in space: It observed the presence of intense bands of radiation that would come to be known as Van Allen belts. America was finally in the game, and the space race — and space exploration — was on.

Ever since *Sputnik* was launched, space exploration and national pride have always been intertwined. The founding of NASA on October 1, 1958, was a direct response to Russia's space achievements — a string of firsts. In addition to launching the first satellite and the first dog, the Russians also sent the first human (Yuri Gagarin, in April 1961) and the first woman (Valentina Tereshkova, June 1963) into space. They also claimed the first person to walk in space (Alexei Leonov, March 1965).

Robotic exploration of the solar system is also replete with Russian firsts, from robotic investigations and landings on the Moon to the first attempted probes of Mars. Like so many others who followed, however, the Russians lost a number of missions to the Mars Ghoul (see page 79). When it came to successful landings on Mars, the American tortoise finally pulled ahead of the Russian hare.

NOVEMBER 16, 1963: Wernher von Braun (left) explains the workings of a *Saturn I* second-stage booster as President John F. Kennedy points toward the rocket.

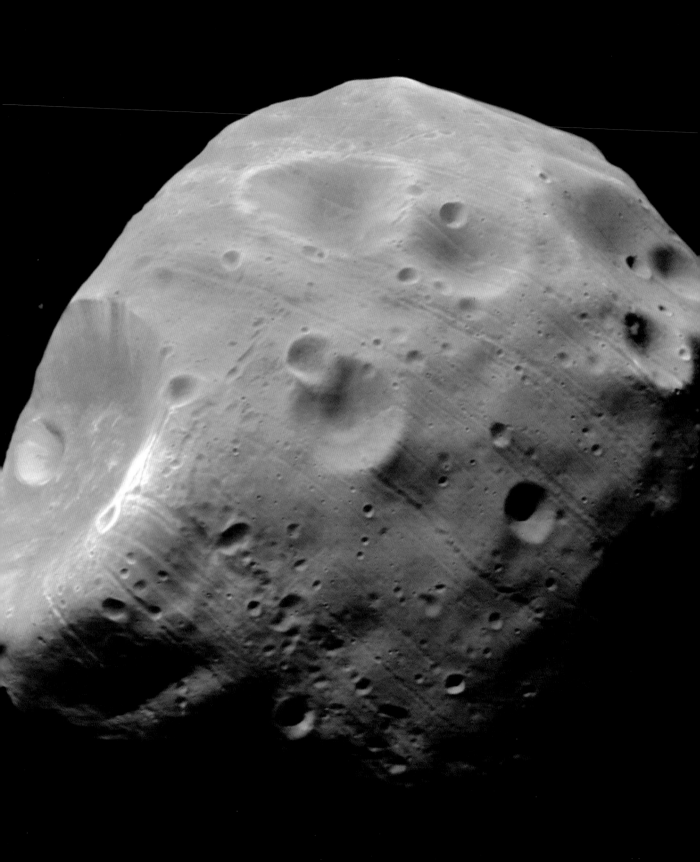

THE MARINER PROGRAM

On July 14, 1965, the first of what Carl Sagan would term humanity's "robot emissaries" transmitted our first close-up views of the Red Planet.

Mars at last!

The eager anticipation with which the pictures were awaited was matched only by the agonizing disappointment with which they were received. At first glance — and the *Mariner IV* mission was exactly that — Mars seemed boring and dull. No canals, no dying races, no cities; just deserts, craters, and hills.

Two more *Mariner* missions, *VI* and *VII*, each made a successful fly-by and relayed to Earth 198 additional pictures. When *Mariner IX* arrived on November 14, 1971, a dust storm cloaked the Red Planet. But this *Mariner* mission was not a flyby: *Mariner IX* successfully entered orbit around Mars and then waited for the storm to subside. Three days later, the storm had not abated, and the Russians arrived with their *Mars 2* mission, followed shortly thereafter by their *Mars 3* mission. Both Russian missions were more ambitious than *Mariner IX*, for they each involved an orbiter and a surface lander. Each lander was equipped with a small, tethered rover that could crawl about 50 feet (15 meters) away from the lander. At least that was the plan.

Mars 2 arrived first. The lander module separated from the orbiter on schedule but entered the Martian atmosphere too steeply: It crashed and burned in the Martian desert — a total loss. Meanwhile, overhead, the *Mars 2* orbiter operated successfully for 362 highly elliptical orbits, snapping pictures and taking measurements.

Mars 3 arrived next. After a successful release of its lander, the orbiter fired its main engine for orbital insertion but then shut down prematurely, most likely because of a fuel shortage caused by leaks. Data flowed from the spacecraft for 20 chaotic orbits before the link shut down.

The lander from *Mars 3* hardly fared any better. The data initially indicated a soft touchdown — which would have made it the first successful landing on Mars. The trouble was that it settled into the maelstrom of the raging dust storm. For 20 precious minutes, signals received by ground controllers back on Earth showed that all systems appeared to be working but the signal was soon lost. Somewhere between the howling winds on Mars and Earth lay the answer, but the loss of the *Mars 3* lander was generally attributed to the dust storm.

Meanwhile, as the storm raged below, *Mariner IX* sent back stunning photos of the Martian moon Phobos and further calibrated the orbits of both Martian moons. About a month after the storm arrived, it cleared, and *Mariner IX* focused on the planet itself, photographing the surface of Mars in unprecedented detail. *Mariner IX* photographed nearly 80 percent of the Martian surface. For the first time researchers began to get a sense of the geological grandeur awaiting further exploration. The largest volcano of any kind in our solar system — it would take three Mount Everests stacked to equal its height — was seen for the first time. Perhaps even more impressive was the discovery of a gigantic canyon which, on Earth, would reach from Gibraltar to Moscow and now bears the name Valles Marineris (Valley of the Mariners). *Mariner IX* transmitted a total of 7,329 images before it closed down just shy of a year after it had entered Martian orbit.

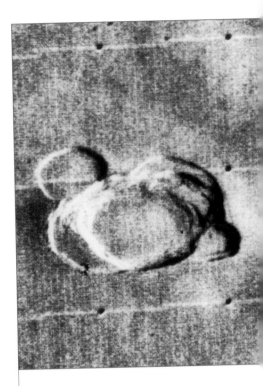

Mariner IX recorded 7,329 images, revealing for the first time many of the planet's features, including this November 1971 image of the largest volcano in the solar system, Olympus Mons.

THE MARS GHOUL

The history of space exploration is peppered with the wrecks of Mars-bound spacecraft. So many that some scientists at the Jet Propulsion Lab have begun to joke about a Great Galactic Ghoul, lurking in the vicinity of Mars, ready to either damage or devour any trespassers from Earth. The Ghoul was first named in 1997 by *Time* magazine reporter Donald Neff. Ghoul talk re-emerged after Japan's 1998 *Nozomi* mission failed, followed by the loss of NASA's *Mars Climate Orbiter, Mars Polar Lander,* and Deep Space 2's penetrator.

Out of 37 launch attempts directed at Mars, only 18 have actually reached the Red Planet. Eleven of these missions were attempts to land, but only 6 have succeeded in transmitting data back to Earth. The Ghoul seems to have it in for Russian spacecraft in particular. To date there has not been one fully successful Russian probe sent to Mars. The only Russian probe that made it to the ground, *Mars 3,* worked for only twenty minutes before it too conked out. The Europeans met the Ghoul when they lost *Beagle 2* in 2003.

A close look at the individual cases reveals the likeliest cause of failure to lie in human error ranging from the purely accidental to the just plain dumb. But Mars is a challenge: It is half the size of Earth and very far away. Landing there requires extraordinary precision and a bit of luck. An element of risk is inherent in the adventure. Joking about the Ghoul is one way to ease the pain of failure.

VIKING

In the main salon of the National Air and Space Museum, in Washington, D.C., in close proximity to the *Apollo 11* command module and the Wright brothers' original *Flyer*, visitors come upon the *Viking* lander. All the items on display in this salon are the original artifacts with the single exception of the *Viking* lander, which is a clone — the actual lander is still on Mars.

So why does *Viking* merit a place among these aerospace icons? The *Viking* mission marked the first time humans actually searched another world for signs of life. The Russians tried, they got there first, but their equipment failed. The British tried in 2003, but their equipment failed as well. *Viking* worked flawlessly, but it has left us with an enigma.

The *Viking* series actually included two orbiters and two landers. Both landers touched down in the northern hemisphere of Mars. The *Viking 1* lander settled in the Chryse Planitia region on July 20, 1976, and the *Viking 2* lander descended farther north, in the Utopia Planitia region on August 7, 1976.

The first image from the surface of Mars was a view of one of the *Viking 1* lander's Frisbee-like footpads on the Martian soil. The photo was taken immediately after the landing because NASA scientists believed that the Soviet landers had sunk into soft sand and they feared a similar result. Some sand *did* settle into the shallow bowls of the footpads, but the photo confirmed that the ground was solid enough to hold the 1,300-pound (600-kilogram) weight of the lander. Then came those marvelous panoramas. For the first time, we could see the horizon of Mars from the ground. What a splendid and desolate place Mars seemed to be!

THE WATERSHED

After the *Viking* mission, the years up to 1996 can accurately be described as lean years for Mars exploration. The Russians sent two missions to Mars's moon Phobos in 1988; both failed. On July 20, 1989, the twentieth anniversary of the *Apollo 11* moon landing, President George H. W. Bush proposed a new Space Exploration Initiative with the goal of landing humans on Mars by the year 2020. But the political will to follow through on this initiative did not exist, and by the time President Bill Clinton took office, it had been lost.

ABOVE: *Viking*'s footpad as seen in this first image from the surface of the Red Planet taken on July 20, 1976.

OPPOSITE PAGE: Gullies in a crater wall at Terra Sirenum as seen by *Mars Reconnaissance Orbiter* in 2006.

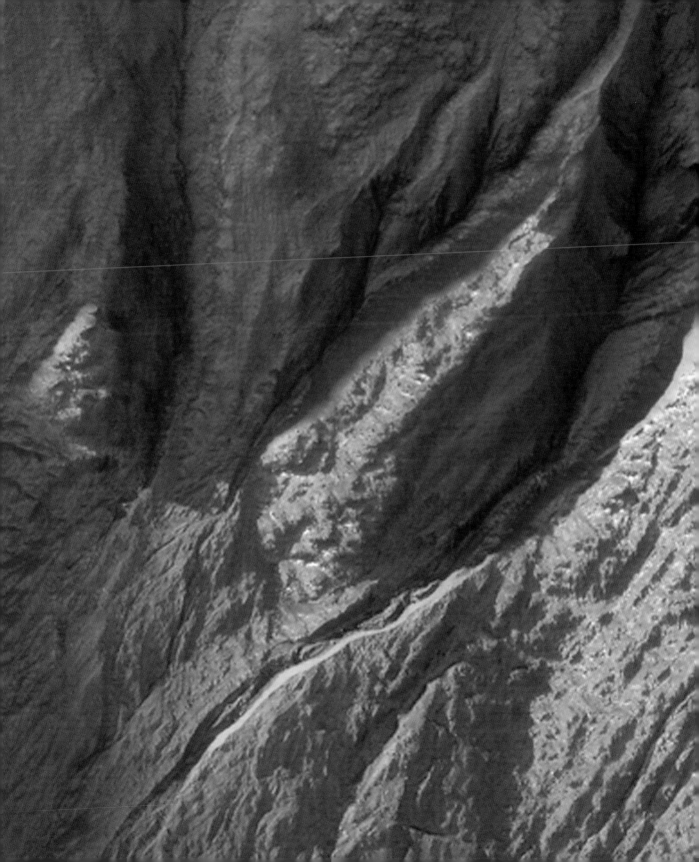

Robotic exploration continued. In 1992, *Mars Observer*, a satellite designed to monitor Martian weather, was lost just prior to orbital insertion. Although the cause of its loss was never fully determined, it is believed that the spacecraft's fuel lines leaked during preorbit pressurization, causing a catastrophic explosion.

Things began to turn around in 1996. The Russians tried to reach Mars that year with a robotic probe based on their earlier Phobos designs but with a different instrument package that featured a ground penetrator. A four-stage Proton rocket was used, and all four stages fired successfully, putting the spacecraft into a planned parking orbit. The fourth stage, with the probe still attached, was to fire a second time to start the probe on its interplanetary cruise, but the second ignition failed, leaving the spacecraft trapped in Earth orbit. Ground controllers successfully undocked the probe from the errant booster, but when they fired its main engine, the probe inexplicably dove into Earth's atmosphere where it burned up.

Meanwhile, NASA's *Mars Global Surveyor (MGS)* was launched on November 7, 1996, and successfully reached Mars nearly a year later. Essentially a replacement for the lost *Mars Observer, MGS* became the longest-operating satellite ever sent into Martian orbit: It remained in continuous operation for more than a decade. It provided the most detailed images of Martian topographical features recorded to date, found gullies carved by water, tracked the formation of Martian dust storms, and monitored Martian weather. Contact with *MGS* was lost on November 2, 2006. A weak signal suggests that the spacecraft is in safe mode, which is a kind of electronic hibernation. At the time of this writing, the reason for the loss of contact is not known, but NASA is trying to reestablish communications.

When *MGS* was launched, it was already overshadowed by a

CONTACT

storm of another sort. What is perhaps the fiercest battle over the question of life on Mars had begun unfolding on Earth.

In August 7, 1996, a strange, sensational story about a rock was the subject of a White House press briefing. President Clinton, along with his science adviser Jack Gibbons and NASA chief Daniel Goldin, announced that a meteor, designated Allan Hills 84001, recovered from Antarctica in 1984, contained evidence of fossilized microorganisms from Mars. David McKay, a scientist from the Johnson Space Center, noted that it contained orange carbonate grains indicating that the meteor had once been immersed in water. Most compelling, however, were the numerous mineralized forms that resembled Earth-like bacteria. McKay contended that while any single piece of evidence could be explained by abiotic processes, collectively the evidence could mean only one thing: microbes on Mars. This argument remains controversial, however, and has since been disputed by many observers, the leading critic being William Schopff, a geology expert from the University of California at Los Angeles. (This controversy is explored in greater detail in Chapter 6.)

And yet, even though lost in dust raised by scientific contention, this stone that fell from the sky has revived our desire to reach the Red Planet as much as anything else.

In December 1996, NASA launched its first rover mission, *Mars Pathfinder*, which landed in Ares Valles in the Chryse Planitia region of Mars on July 4, 1997. Its rover was dubbed *Sojourner* in honor of the abolitionist Sojourner Truth. Although this mission was ostensibly to analyze the Martian atmosphere, climate, rocks, and soil, its greater purpose was to serve as an engineering test bed for new rover features such as an airbag touchdown and an automated obstacle avoidance system. It provided a low-cost, proof-of-concept demonstration for NASA administrator

Daniel Goldin's "faster, cheaper, better" development principle. But *Mars Pathfinder*'s most significant effect may have been on the Internet. On July 9, five days after *Pathfinder* landed, NASA reported more than 47 million hits at its *Pathfinder* Web site — a remarkable achievement in 1997. The daily average for all NASA's Web sites in late 2006 was 8.3 million hits.

THE GHOUL STRIKES AGAIN

Following the success of *Global Surveyor* and *Pathfinder*, the exploration of Mars was set back by a string of failed missions. The Japanese *Nozomi* orbiter (part of an international joint venture including Canada, Germany, France, Sweden, and Japan) was intended to study the upper atmosphere of Mars, but the probe burned too much fuel trying to escape Earth. Heroic efforts to save the mission with a number of gravity-assist maneuvers past Earth and the Moon put the probe within 620 miles (1,000 kilometers) of Mars, but there was not enough fuel for orbital insertion. *Nozomi* was parked in a new heliocentric orbit, a move designed to keep it from crashing and contaminating the Martian surface with bacteria from Earth.

This was followed by NASA's own mishaps: the crashing of the *Mars Polar Lander* and the loss of *Mars Climate Orbiter*. The MCO was also lost while attempting orbital insertion. A goof had been made in the conversion of English to metric units, causing the spacecraft to enter the Martian atmosphere at a much lower altitude than was called for by its flight profile; it burned like a meteor in the Martian sky.

The *Mars Polar Lander* almost made it, but a software bug resulted in the misinterpretation of vibrations caused by deployment of the lander's legs, signaling the engines to shut off prema-

The long, rocky road traveled by the rover *Opportunity* along the western edge of the Erebus Crater is photographed February 26, 2006.

turely when the spacecraft was a still about 130 feet (40 meters) above the ground. All contact was lost.

In 2003, the European Space Agency launched the *Mars Express Orbiter* from the Baikonur Cosmodrome in Kazakhstan. Attached to the orbiter was a small lander named *Beagle 2* in honor of the ship that carried Charles Darwin on his voyage of discovery. This lander represented the first attempt to find life on Mars since the *Viking* programs, but communication was lost after the lander descended into the Martian atmosphere. What went wrong is not known. The orbiter, however, was successfully placed into Martian orbit and continues to send back stunning images to this day.

SPIRIT AND OPPORTUNITY

But there is nothing like success, and the rovers *Spirit* and *Opportunity* have been operating on Mars continuously since January 2004. In early 2007, both rovers were still going strong.

Spirit, or *Mars Exploration Rover-A (MER-A)*, landed in a dried lake bed, Gusev Crater, named for the Russian astronomer Matvei Gusev, about 14 degrees south of the equator. *Spirit*'s twin, *Opportunity (MER-B)*, landed on the opposite side of the planet on the Meridiani Planum (Plateau of the Meridian).

These rovers found that Mars had once had a series of shallow seas, as indicated by the abundance of hematite, a mineral that forms only in the presence of water. So what happened to the water? Is it locked up in the soil as ground ice? Did it evaporate into space as the Martian atmosphere thinned? At the time of this writing, both are in hibernation, awaiting the end of their second Martian winter. *Spirit* appears to have one wheel stuck, but the little rover seems to be chugging right along.

ROVERS

Jubilant mission planners described the rover *Opportunity's* soft landing in the center of Eagle Crater as an interplanetary "hole-in-one." The rover's tracks can be seen in this February 17, 2004 image of the rover's landing platform and deflated airbags.

THE SITUATION TODAY

In the late winter of 2006, scientists using the now-defunct *Mars Global Surveyor* discovered a new run-off gully that had formed between December 2001 and April 2005. These are the dates of two photos taken of the northwest wall of a crater in Terra Sirenium region of Mars. A similar gully has recently formed in a crater in the Centauri Montes region. There is some debate over the cause of these gullies — one school of thought is that they are caused by landslides, but a more obvious and far more tantalizing theory is that they were caused by pockets of newly exposed water. The source of this water could be aquifers, snow packs, or ground ice. If this evidence for subsurface reservoirs pans out, the gullies would be likely places to find more water and perhaps indigenous microbes.

The shoals of Mars are peppered with wrecks from nearly two dozen failed missions. Two robot rovers hibernate, waiting for winter's thaw, while a host of robot spacecraft pass through the sky making their daily rounds. Mars has surprised us by both what it lacks and what it has.

Few would have believed, in the last years of the 1960s, that human exploration of the Moon would end with the *Apollo* mission, and that in the early years of the next millennium NASA's budget would be so modest that the space agency would be struggling just to maintain a presence in low Earth orbit. The mighty Soviet space program, once a sure bet to put the first humans on Mars, is scarcely a shadow of its former self. Across the gulf of space, about 40 years after receiving its first robot visitors from Earth, Mars remains cool and unsympathetic, regarded by Earth with envious eyes, a world that is vast and mysterious and not yet touched by human hands.

This *Mars Global Surveyor* image from September 2006 shows part of a north polar dune field during the Martian summer. Usually polar dunes are covered with frost, but in late spring and summer the dark, windblown sand becomes fully exposed. This view, located near 79.8°N, 127.1°W, covers an area approximately 1 mile (1.5 kilometers) wide.

GETTING
THERE

"ANY ASTRONAUT, YOU SCRATCH OUR SKIN AND YOU'LL FIND MARS BLOOD
FLOWING UNDERNEATH. THAT'S HOW MUCH I CARE ABOUT IT."
— Jeffrey Hoffman, astronaut, ABC News interview, January 15, 2004

January 16, 2029. Ares, the Greek name for Mars, is now also
the name of the mightiest rocket to be flown since the glory days
of the *Saturn V* Moon Missions. When *Ares* begins its 10-day
rollout from the Vehicle Assembly Building (VAB) at the
Kennedy Space Center in Florida, news footage shows a towering giant emerging into the light. America has at last reclaimed
the heavy-lift capacity that was lost when it abandoned the
Apollo program.

The dramatic December 9, 2006, night launch of the space shuttle *Discovery*.

This will be the first of the two *Ares* launches, each of which will carry a payload (*Shirase A* and *Shirase B*) into space. On the evening before launch, the rocket can be seen from a great distance. It stands in formidable silhouette at the apex of a pyramid of light beams, the pointed nose cones of the main rocket and the two solid booster rockets mounted on its sides suggesting a triptych of Gothic arches.

A short ride on the VIP bus toward Launch Pad 39B brings a lucky few visitors close to this mighty giant.

Security is reassuringly tight. But it is worth it, for being this close to *Ares* is to stand humbled in the presence of a technological almighty. The feeling that *this* is what the world should be doing, that *this* is the embodiment of all that is good about the human race, sanctifies the ground here. Behold at last the hand of *Homo sapiens* reaching for the stars.

The massive test vehicle for the *Saturn V* rocket is rolled to the Kennedy Space Center's Launch Pad Complex 39A in June 1966, from the Vehicle Assembly Building on the left.

Now back to the VIP viewing stand across the Banana River. The clock ticks off the seconds as launch time approaches.

"T-minus two minutes and we are now on a scheduled hold in the countdown."

The voice of the capsule communicator (CAPCOM) booms across the VIP stand and runs down a long, well-rehearsed checklist of cryptic acronyms. The voices of different men and women snap their replies in a cadence that only the cognoscenti can appreciate:

"EMS?"

"Go."

"COMDEV?"

"Go."

"COMEEKO?"

"Go."

"SATCOM?"

"Go."

CAPCOM resumes the countdown.

"We are transferring to internal power at this time. *Ares* is now running off its onboard fuel cells. Coming up on a go for autosequence start. And we have a go for autosequence start. *Ares*'s onboard computers have primary control of all the vehicle's critical functions. T-minus seventeen seconds and counting. Fifteen. Twelve. Eleven. Ten. Nine. Eight. Seven. Six. Go for main engines start. Main engines start."

"Two. One. Booster ignition and liftoff of the *Ares V Shirase A* rocket, blazing the trail for our future on Mars."

Ares lifts slowly and rises on a pillar of flames. It seems to take an eternity to rise above the gantry, but it does, lighting up the entire sky in dazzling amber hues.

"Houston now controlling the flight of *Ares*."

The rocket is hardly clear of the tower when it starts to tilt slightly and roll over about 15 degrees.

Now the rocket appears to accelerate and heads downrange for the 8-$^{1}/_{2}$-minute climb to orbit.

Viewers watching the whole spectacle from the VIP bleachers at Banana River stand in dumbfounded amazement. The sound and the pressure of the launch can be imitated by drumming one's fingers lightly across the chest. In a flash *Ares* is gone, leaving behind only a boiling cloud of steam and hydrochloric acid to mark its departure. A giant television monitor next to the countdown clock shows pictures of the rocket, which is now far beyond the range of the human eye.

"*Ares* is already 3-$^{1}/_{2}$ miles [5.6 kilometers] in altitude and 1-$^{1}/_{2}$ miles [2.4 kilometers] downrange, traveling almost 750 miles [1,207 kilometers] per hour. Everything looking good on the bird."

The sheer beauty of spaceflight can sometimes mask its perils. Here, with the Moon beckoning on January 16, 2003, the space shuttle *Columbia* arcs into the night sky. This was *Columbia*'s last flight.

"Fifty-seven seconds into the flight. Engines beginning to rev up. Standing by for the throttle-up call from CAPCOM."

"Copy. Now 1 minute, 47 seconds into the flight, 22 miles in altitude, 18 miles downrange, traveling 2,600 miles an hour [35 kilometers in altitude, 29 kilometers downrange, traveling at 4,200 kilometers per hour]. Standing by for solid rocket booster separation."

"SRB separation confirmed. Guidance now converging. *Ares* onboard computers commanding the main engine nozzles to swivel, aiming the rocket for its precise target in space for main engine cutoff."

If the CAPCOM procedure sounds a little familiar, it should. *Ares V* is derived from space-shuttle components. The advantage of *Ares* over the shuttle is that it nearly restores the heavy-lift capacity of the *Saturn V*. The *Saturn V* could lift 260,000 pounds (118,000 kilograms) into low Earth orbit, while the shuttle could lift only 62,000 pounds (28,000 kilograms). The *Ares V* can lift as much as 240,000 pounds (108,000 kilograms), slightly less than the old *Saturn V*.

As *Ares V Shirase A* makes its way into space, another *Ares* rocket, *Ares V Shirase B*, is ready to launch the second half of the *Shirase* payload: it's so large and heavy that two rockets are needed. The payloads will rendezvous in low Earth orbit and then push on to Mars.

The *Ares V* rocket, on the right, towers over its predecessor, the space shuttle. *Ares* is a fusion of shuttle and *Apollo*-era concepts. Like the shuttle, *Ares* uses two solid rocket boosters strapped to its sides and has almost as much lift capability as the old *Saturn V*.

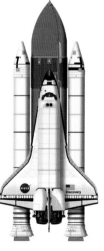

Shirase's manifest includes a pair of surface exploration vehicles (SEVs) — unpressurized three-seat trucks that the *Terra Nova* crew will use for ground transportation. Also on the manifest is a small nuclear reactor based on a design derived from the Department of Energy's Lawrence Livermore National Laboratory.

This small, sealed, transportable, autonomous reactor (SSTAR) can deliver 10 megawatts of electricity for 30 years without needing any additional fuel. Since the unit is entirely self-contained inside a hermetically sealed cask, there is no messy radioactive spent fuel to worry about; instead, the entire unit will be replaced. The old 2007 version of the SSTAR weighed approximately 200 tons (180 metric tons), but this fourth-generation unit bound for Mars is an order of magnitude less, at only 20 tons (18 metric tons).

A schematic of the six-wheeled surface exploration vehicle (SEV) rover. Two electric rovers will be employed for exploration and for carrying equipment and samples.

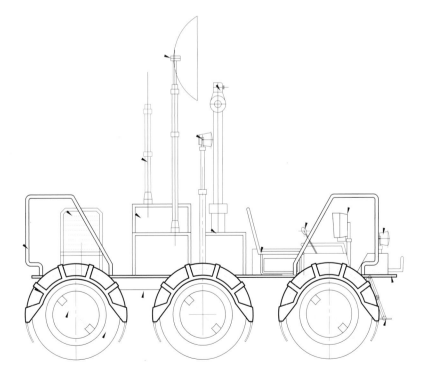

The last major item on the list is a wireline drill system. Wirelines are typically used by the oil and gas industries to lower equipment into a well. The metal threads of cable wire are wrapped around an inner core of insulated power lines and data cables that run to equipment located at the end of the cable behind a diamond drill bit. These data cables provide telemetry and communications between equipment at the end of the cable and the astronauts on the surface. On Mars the wireline drill is carried to the site designated for drilling by one of the SEVs before the *Terra Nova* crew lands.

As in the case of the *Shirase* mission, two rocket flights are needed to launch the *Atlantis* surface habitat, the module that our astronauts will call home when they are on Martian turf. The *Ares V Atlantis A* and *B* are both set to lift off from the Kennedy Space Center 2 weeks after the *Shirase* launches. The two halves of the *Atlantis* module will rendezvous in low Earth orbit and then push off as a single module toward a rendezvous with Mars. *Atlantis* features bunks for the crew, a dual-chamber airlock, a common room, showers, a workout facility, and a kitchenette.

The Russian Space Agency is responsible for building and sending the MADV *Gagarin* in two launches from Baikonur. The Russians will use their *Energia* rocket (a heavy-lift monster that, prior to this Mars mission, had been used only twice, first to launch the sole flight of the Russian space shuttle *Buran* and then to deploy the Polyus anti-satellite weapon, a vehicle that carried a recoilless cannon for use against American Star Wars platforms, and communication and spy satellites). Like the *Shirase* and *Atlantis* payloads, the *Gagarin* module will be sent in two pieces into low Earth orbit where each will dock and then proceed to Mars as a single unit. *Gagarin* will then orbit Mars and await rendezvous with the main *Terra Nova* crew transit vehicle.

RENDEZVOUS

An animated rendering of setting up a Mars Base: Most of the equipment needed for the first sojourn on the Red Planet is sent in advance of *Terra Nova*. In the top panel, *Atlantis* — the crew's equivalent of a bed-and-breakfast on Mars — comes in for a landing, observed by cameras on one of the rovers from the already-deployed *Shirase* module. In the center panel, the second of the two rovers is deployed in autonomous mode from *Shirase* to help prepare for the guests. In the third panel, a cable is strung from *Shirase* to *Atlantis* by the automated rover.

TRAVEL PLANS

The NTR cargo vehicles carrying the *Atlantis* surface hab, the SEVs, the *Gagarin* lander, and other supplies have already been launched toward Mars. Now the *Terra Nova* crew transit vehicle, the world's first manned interplanetary spacecraft, is on its way to join them. This spacecraft has three primary modules: a hard-shell crew module (room for six) dubbed the *trans hab*; a nuclear reactor which provides electrical power to the ship and serves as a nuclear rocket engine; and four modular fuel tanks, all arrayed along a single truss. Also tucked into the truss is the earth return capsule (ERC), which is a modified *Apollo*-esque *Orion*-class crew exploration vehicle (CEV) developed in the latter half of the first decade of the twenty-first century. The ERC will detach from *Terra Nova* as it flies past Earth and will bring the crew back home. *Terra Nova* will then enter a heliocentric parking, or stationary, orbit.

Terra Nova, after having been assembled in low Earth orbit, prepares for its sortie into interplanetary space.

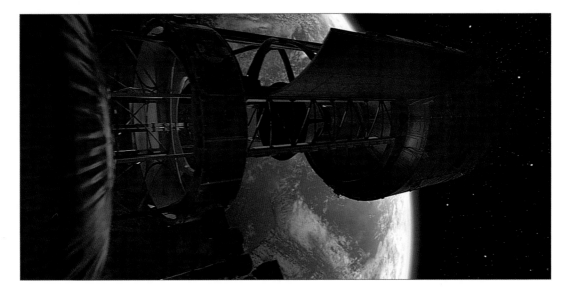

ZUBRIN'S BIG IDEA

"Are we still a nation of pioneers?"
— Robert Zubrin, *The Case for Mars*, 1996

Most schemes for a human mission to Mars contain elements of the Mars Design Reference (MDR) plan. The number of rocket launches varies from one plan to the next, depending on the selection of trajectory and rocket. (For example, the *Energia* rocket can lift 110 tons (100 metric tons) into low Earth orbit, but in its Vulkan configuration it can lift 193 tons (175 metric tons). The MDR, in turn, is derived from a mission template first put forward by Robert Zubrin, a former Lockheed-Martin engineer. It is unlike all previous Mars mission plans, which Zubrin characterized as the Battlestar Galactica approach. Those missions were designed to bring everything with you. Zubrin's big idea was that a Mars expedition should plan to live off the land, exploiting as many in-situ resources as possible. A robot factory, for example, could synthesize the fuel needed for the trip back home from Martian resources before humans ever left Earth. This would eliminate the burden of hefting that fuel to Mars.

There are some notable differences between the events described in this book and the plans Zubrin has drawn up. In Zubrin's plan, the astronauts would live on Mars for a year and a half, rather than the 60 days described in these pages. They would take a direct Mars-trans flight that would occur when Mars was at opposition. It would not include a flyby of the planet Venus as described here. Zubrin has also proposed that the rovers be powered with fuel produced by a robotic fuel factory, and not by electricity, citing the higher power-to-weight ratio that gas-engines can provide. In defense of the electric cars, however, it should be noted that the power-to-weight ratio is improving due to significant advances in battery technology.

Zubrin's epiphany came when he was studying the history of polar Arctic and Antarctic exploration: The expeditions that lived off the land stood the best chance of success. Zubrin's scheme for going to Mars is the antithesis of von Braun's *Das Marsprojekt*. It's significantly less expensive, relies on existing technology, and builds infrastructure along the way.

January 30, 2030. The *Orion* CEV *Columbia* has docked with *Terra Nova*, and the crew is now making last-minute preparations for their long voyage. There are few human activities for which there is no precedent, but this journey is one of them. The outbound portion of the trip will take 11 months and will include a Venus flyby.

The astronauts will spend 62 days in the Dao Valles region of the Hellas Basin in the southern hemisphere of Mars. The return flight will take 6-$\frac{1}{2}$ months.

At the beginning of the journey, while *Terra Nova* is still in low Earth orbit, communications with CAPCOM are in real time. As *Terra Nova* gets farther away from Earth and closer to Mars, there will be a lag in communications. These delays will eventually become so long that they will inhibit interactive dialogue.

"We are coming up on a go for *Terra Nova*'s autosequence start. And we have a go for autosequence start. *Terra Nova*'s onboard computers have primary control of all the vehicle's critical functions. T-minus seven seconds and counting. Three, two. Go for main engines start."

Terra Nova trembles a little as her vector changes.

"Main engines start. Burn is nominal."

"Copy."

"*Terra Nova*'s main engine burn is nominal."

The first voyage by humans to the Red Planet is now underway.

2030:01:30

HAZARDS OF
LONG-DURATION SPACEFLIGHT

As the members of *Terra Nova*'s crew make their way toward their historic rendezvous with the Red Planet, what risks do they face? In March 2004, astronauts on board the *International Space Station* suddenly found themselves confronted by an ominous development.

"It's not life-threatening material. The situation is stable now," the information officer told the press gathered at Johnson Space Center in Houston. It seemed almost as though he was trying to reassure himself as much as the press corps. The announcement followed the discovery of traces of noxious gas inside the *ISS*. "We have a leak," the crew told Houston. "We haven't pinpointed the source yet, but there's potassium hydroxide coming out of the ventilation system."

Ground controllers immediately declared a spacecraft emergency which allowed them to commandeer any communications satellite in the sky they needed to maintain live, real-time communication, but they stopped short of ordering the crew to rush to the *Soyuz* lifeboat. The leaking gas was identified as a Tox-2-level material — irritating but not lethal.

Environmental conditions inside *Terra Nova* are monitored constantly by both humans and computers.

The crew, trained to respond to emergencies like smoke and fire, toxic leaks, and rapid decompression, shut down the ventilation system, donned gloves and surgical masks, and traced the leak back to their Elektron unit inside the Russian-built *Zvezda* module. The Elektron, also Russian-built, makes breathable oxygen through electrolysis — the separation of hydrogen atoms from oxygen atoms in water. The water itself is nonpotable, and though most of this water is brought up by the space shuttle, some of it is recovered from humidity in the air and from recycled urine. The hydrogen is then vented outside the station, and the oxygen is circulated back into the air. The smoke emanated from a rubber seal on the Elektron's interior plumbing.

The Elektron unit is finicky. It was the source of a similar scare in the same year, but it's not the only possible source of troublesome gases. Two years earlier, mysterious odors sent the *ISS* astronauts fleeing into the Russian segment of the station. These odors wafted from the U.S.-built Quest airlock, where extravehicular activity (EVA) suits are stored. A pungent mildew had settled into a system that cleanses and recharges air scrubbers inside U.S. spacesuits.

Toxicity ratings for all materials found on *Terra Nova* are based on NASA's classifications for the *ISS*, as shown in the table below.

TOXICITY HAZARD RATINGS FOR THE *ISS*	
NONHAZARD	TOX-1
CRITICAL HAZARD	TOX-2
CONTAINABLE CATASTROPHIC CONTACT HAZARD	TOX-3
CONTAINABLE CATASTROPHIC SYSTEMIC HAZARD	TOX-4
NONCONTAINABLE CATASTROPHIC SYSTEMIC OR CONTACT HAZARD	TOX-5

The *International Space Station* as seen from the space shuttle *Discovery* on the first "return to flight" following the loss of *Columbia*. The station has been continuously occupied since November 2, 2000, and will be completed by 2010.

REHEARSING IN SPACE

The *International Space Station* orbits at an average altitude of 220 miles (360 kilometers). In comparison, the *Hubble Space Telescope* averages an altitude of 375 miles (600 kilometers), which is pretty close to the operational ceiling of the shuttle. With the addition of a new set of solar panels in late 2006, the *ISS* has become brighter than the planet Jupiter and can be seen as a pale yellow star moving steadily across the night sky. When the station is completed, it will outshine Venus to become the third-brightest object in the sky after the Sun and the Moon.

The *ISS* has been the object of searing criticism over the years, derided as a political pork barrel that diverted funds away from genuine science missions to create make-work projects for white-collar engineers. It began as the centerpiece of President Ronald Reagan's manned space program, but with the end of the Cold War, it became the focus of international cooperation. Participants include NASA, the Russian Space Agency (RKA), ESA, JAXA, CSA, and the Brazilian Space Agency (AEB).

At first glance, the *ISS* seems to provide an ideal test bed for the kind of long-duration spaceflight that a trip to Mars would entail. There are similarities, but the differences are significant.

In the case of the gas leak described earlier, thanks to international cooperation, ground controllers were able to commandeer any available satellite to establish and maintain constant two-way conversation. The distance involved in the voyage to Mars eliminates the possibility of real-time intervention from the ground: the astronauts will have to wait for answers. This is because of the finite speed of light — 186,000 miles per second (300,000 kilometers per second). Radio waves, microwaves, visible light, X-rays, and infrared light are all different frequencies of the same thing — electromagnetism — and consequently are constrained by the top speed at which light can travel. If, for example, two objects are 372,000 miles (600,000 kilometers) away from each other, it will take light 2 seconds to travel the distance between them. Radio communication from one object to the second object will take 2 seconds, and then there will be another 2 seconds for the reply — for a total lag of about 4 seconds. The lag each way between Earth and Mars depends on the position of Mars in relation to Earth. When Mars is at opposition, the lag is about 6.5 minutes one way. When Mars is at superior conjunction (its greatest distance from Earth), the lag can be as much as 44 minutes each way.

The rhythm of communications between the spacecraft and CAPCOM will gradually become disrupted as the Mars-bound vehicle leaves the vicinity of Earth. This means that conversations and instant messaging will eventually become impossible. Other forms of communication, such as e-mail and voice mail, will work just fine because they require no direct interaction.

Another fundamental difference between an *ISS* and a Mars mission is that astronauts aboard the *ISS* can bail out whenever they want. Note again that in the leaking-gas scenario, ground controllers had the option of ordering the crew into the safety of one of the *Soyuz* lifeboats for a quick, safe return to Earth. The Mars-bound *Terra Nova*, however, can't simply turn around and fly back to Earth. As the vehicle leaves Earth orbit, the option to abort diminishes with increasing distance. After the sixtieth day of the mission, this option disappears. For better or worse, the crew is stuck for the duration. For a Mars-bound crew — unlike the crews of the space shuttle — there is no chance of rescue. They will be like the ancient explorers from the age of discovery: completely on their own.

THE CABBAGE PATCH

From a biologist's point of view, the *ISS* is a wonderful laboratory for studying the growth and development of plants and some animals in space. Experiments like the Ames Research Center's Botanical Production System (BPS) look at the dangers of long-term exposure to microgravity and ways to reduce the risks. The BPS, in particular, was designed to study vegetation in a weightless habitat and to evaluate automated growing equipment designed for a long spaceflight. It's essentially a greenhouse in a box that lets ground controllers monitor the photosynthesis and

Botany will play a vital role in nurturing both mind and body during the long flight to Mars. The greenhouse used on *Terra Nova* was originally developed and tested on the *ISS*.

health of the vegetation. The plants are grown hydroponically: Their roots are immersed in water and fed by nutrients in the water. The plant takes in the nutrients, and the water is recycled. The lighting, heat, and air are tightly controlled. On board the Mars-bound *Terra Nova*, more than a dozen BPS-derived gardens have been installed.

Studies have shown that seeds grown from plants in space are smaller and of poorer quality than their Earth-side counterparts. Their composition is subtly different: They have larger-than-normal amounts of starch when they should have been storing proteins and lipids. Clearly this may be a problem for the crew of a Mars-bound spacecraft. The value of raising crops en route cannot be overestimated. Apart from their value as fresh food, plants scrub the air by removing carbon dioxide, making oxygen, and purifying water.

Plants also can be a source of stress release, as Michael Foale discovered during his tenure on *Mir*. The flowers of *Brassica rapa* plants, like those taken on board the *ISS* in 2002, are pollinated by bees. Foale used the body of a dried bee on the end of a toothpick to pollinate *Mir*'s garden by hand. He looked so peaceful while doing this that fellow astronaut Storey Musgrave nicknamed him "Foale the Farmer."

MIND GAMES

A journey to Mars is teeming with risks, but perhaps the greatest is that posed by the travelers themselves. The isolation, boredom, crowding, homesickness, and tensions that arise among crewmates are all potential safety threats. A breakdown in communication between crewmates who speak different languages and who grew up in different cultures also poses a risk. These dangers can be minimized through careful screening and rigorous training before takeoff. Crew members have to be compatible.

Fraternization has to be carefully watched as well. This problem is best captured by the character of Hannibal Lector in the movie, *Silence of the Lambs*, when he famously and succinctly stated, "we covet what we see." The only live human contact during a long voyage to Mars will be the other members of the crew; family and colleagues on the ground will be mere digital images on a screen. The team members will form bonds, but the danger is that someone will feel left out, spurned, or jealous. Something like this appears to have happened in the case of Lisa Nowak in 2007 when she formed an unrequited attachment for another astronaut although, in this case, the emotional breakdown occurred on the ground. Should someone snap during a cruise to Mars, the consequences could be disastrous.

It is also important to provide the right amount of work. In November 1973, the flight crew on *Skylab IV* fell behind in completing their list of tasks and the ground crew chastised them for not keeping up. This resulted in an open rebellion on the ship. The crew's workload was subsequently reduced, but their rebelliousness was not entirely quelled: They defied orders not to photograph Area 51. (The resulting images of this super-secret military facility have never been released.)

"All the necessary conditions to perpetrate a murder are met by locking two men in a cabin of 18 by 20 feet...for two months."
— VALERY RYUMIN, COSMONAUT

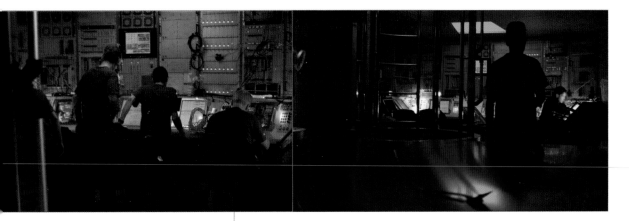

In addition, the assigned work has to be carefully planned so that the super-achieving mentality of astronauts is not under-challenged, either by a paucity of meaningful tasks or a preponderance of menial chores like those associated with housekeeping. Care must be taken to ward off depression and emotional problems. There have been a few cases of depression that have led to in-flight changes in motivation, diet, sleep, and exercise. No matter how thoroughly prepared crews are for long flights, the U.S. and Russian experiences have shown that at least a few astronauts will experience bouts of depression, posing a risk to both individual crew members and the mission. Veteran astronauts have hobbies: Michael Foale gardened; Shannon Lucid read books; Carl Walz played a keyboard.

Sleep deprivation caused by disruption of the circadian rhythm can also invite depression and impede an astronaut's performance. In low Earth orbit, the *ISS* circles the Earth nearly 16 times every day, causing the crew to experience a corresponding number of artificial dawns and sunsets. A trip to Mars will have the opposite problem — no sunrises or sunsets at all, just full sun for as long as the crew is in transit from one planet to the other. The effect will be similar to that of the Midnight Sun known to people who experience an Arctic summer. Strict scheduling and medication can help maintain the circadian rhythm.

NASA has a tradition of playing wake-up songs for its astronauts as a way to make sure their day begins on time. On the evening of November 13, 2005, the two-man crew of the *ISS*, Bill McArthur and Valery Tokarev, were treated to the song "Good Day Sunshine" performed live by former Beatle Paul McCartney. Both the astronauts and the singer were thrilled.

BONE LOSS

Osteoporosis is a nearly invisible but serious threat to astronaut health. The rate of bone loss, as demonstrated by both the *ISS* and the *Mir* programs, is from 1 to 2 percent a month. One *Mir* cosmonaut returned to Earth after a 6-month sojourn with a crippling 20 percent loss of bone density.

The repercussions can be dire. The odds of fracturing weakened bones are greatly enhanced and, once broken, they heal slowly. The chances of developing a painful renal stone may also be increased because of the elevated urine-calcium concentration associated with bone resorption. The only way to eliminate this is through increased hydration.

There is some evidence that cardiac dysrhythmias can result from extended weightlessness. We know that in space blood pools in the torso and head, giving astronauts puffy faces, but relatively little is known about alterations in the electrical pattern that causes the heart to pump blood. There is some evidence that the heart itself undergoes a loss in mass and strength, which jeopardizes the performance of any strenuous task or, for that matter, any excursion to a planetary surface. Dysrhythmias can lead to possibly fatal heart attacks.

"I can't believe that we're actually transmitting to space! This is sensational. I love it."
— PAUL McCARTNEY, FORMER BEATLE

"Flying in space brings many, many exciting events, but I must tell you this ranks at the very top."
— BILL McARTHUR, *ISS* ASTRONAUT

Muscle and bone loss will be an inevitable consequence of the journey, especially if there is no artificial gravity. Astronauts now always adhere to a rigorous regimen of exercise to maintain condition.

The chances of developing allergies and getting sick during a long space-flight are also increased because there is some evidence that the lack of gravity and the increased presence of radiation cause autoimmune deficiencies, particularly involving immune cells like CD4$^+$ (helper) T cells, B cells, NK cells, monocytes/macrophages/dendritic cells, hematopoietic stem cells, and cytokine networks.

There are two ways to prevent these maladies. One is to design the spacecraft so that it can induce artificial gravity through centrifugal force. By spinning *Terra Nova*, the crew members are able to keep their feet firmly on the floor. As of this writing, NASA is looking for volunteers to participate in long-duration experiments involving spin-induced artificial gravity at the Johnson Space Center in Houston, Texas. These tests are necessary because we don't know whether a 1.0-G centrifugal force is an adequate substitute for gravity. The second remedy is vigorous exercise. A number of astronauts and cosmonauts, overcome by weakness, have had to be carried from the capsule at the end of their journeys. But when Valery Polyakov returned from his record-setting 438 days in space in 1994–1995, he walked away under his own power, thanks to an aggressive exercise regimen. On board *Terra Nova*, the crew will be required to exercise 3 hours a day to keep in shape for the rigors that lie ahead of them.

RADIATION

Without the protective atmosphere of Earth and the Van Allen belts, space travelers are vulnerable to radiation from the Sun and to cosmic radiation. All kinds of nasty radiation can be found in the interplanetary medium. Everything from gamma rays and X-rays to microwave radiation and atomic and subatomic particles zing space travelers. Astronauts inside the *ISS* receive about the same amount of radiation in a week to 10 days as people on Earth receive from background radiation in a year. On an interplanetary voyage, astronauts reach the maximum yearly dosage allowed for radiation workers at nuclear power plants on Earth within approximately 2 months. Lifetime limits for radiation exposure have been officially set at higher levels for astronauts than for nuclear-power-station workers — not exactly a perk for the astronaut corps.

On board *Terra Nova*, the entire crew module is shielded, but the crew compartments have additional shielding to reduce the exposure resulting from a solar flare.

Radiation will be an abiding danger during every portion of the journey and on Mars itself. Shielding for radiation, however, can be built discreetly into the structure as shown in this interior sketch of *Atlantis*.

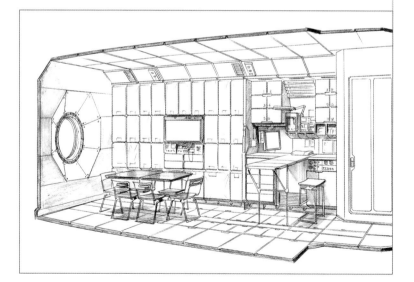

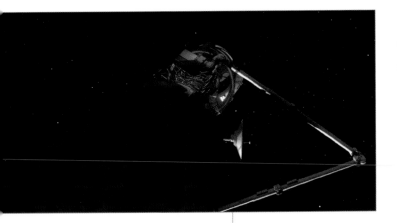

Terra Nova will have a robotic arm much like the Canadarm used on the space shuttle and the remote manipulator arm (RMA) on the *International Space Station*. Operated from within the Trans Hab, the RMA on *Terra Nova* is seen here looking for potential damage to the spacecraft's power unit.

During upheavals on the Sun, radiation levels can soar off the charts. A single coronal mass ejection (CME) can pump enough X-rays into space to fry the electronics of any orbiting satellite. The good news is that interplanetary voyagers may have ample warning of the approach of an incoming CME. This is not the case for gamma ray bursts (GRBs), however, which come from outside our galaxy. By the time NASA's orbiting *Swift* satellite (*Swift* is designed to chart the source and direction of GRB flashes) can detect a GRB, the danger has already arrived.

Perhaps the most vulnerable period for an astronaut is during an extra-vehicular activity (EVA) — a space walk — outside the protective shielding of the spacecraft. Except for timing a space walk so that it occurs on the shadow side of a spacecraft, or armoring EVA suits so that their wearers resemble medieval knights, there is really nothing that can be done to mitigate the radiation risk outside the ship.

The immediate danger from radiation is caused by free radicals (molecules with an unpaired electron) rupturing cell membranes. This rupturing causes the cells to die. If it occurs on a large scale, it can lead to organ failure. The long-term danger is carcinogenesis; the risk of cancer due to mutations caused by damaged DNA is heightened. For interplanetary spaceflight, one solution lies in the use of polyethylene. It turns out that a high-density version of the same polyethylene plastic used to make 2-liter soda bottles and trash-bin liners also makes excellent radiation shielding for spacecraft. Polyethylene has a high hydrogen content, and hydrogen atoms do an excellent job of absorbing and dispersing radiation. Another defense is to feed space travelers a diet high in antioxidants — enzymes or other organic molecules that can counteract the damaging effects of free radicals. This means that foods high in vitamin C, vitamin E, selenium, and carotenoids (such as beta-carotene and lycopene) will be prominently featured on the astronauts' dinner menu.

FIRE

Any fire from which there is no escape is terrifying. In a spacecraft — where the smoke generated by a conflagration involving a multiplicity of plastics and other synthetic materials is bound to be toxic — even a small blaze can be deadly. When flames erupted on board *Mir* in 1997, Jerry Linenger became the first NASA astronaut to fight a fire in space. The blaze was caused by a leak from a lithium perchlorate canister, which is a kind of oxygen candle commonly used on Russian submarines, in mines, and on spacecraft. When lit, it smolders at more than 1,100 degrees Farenheit (600 degrees Celsius) for up to 20 minutes, producing as by-products sodium chloride, iron oxide, and about 6.5 man-hours of breathable oxygen per kilogram of the mixture. On February 23, 1997, lithium perchlorate leaked from one of the lighted canisters, which emitted molten-iron sparks like Fourth of July fireworks. The canister burned for 14 minutes and blocked the escape route to one of the *Soyuz* spacecraft until it was finally extinguished.

The outbreak of fire at any stage during the journey to Mars could very quickly end in catastrophe. On *Terra Nova* an extinguisher is used to put out a fire in some of the circuit panels in the flight control room.

How a fire burns in space is a subject in need of further study. Soot from a fire in space, for example, is larger than soot on Earth. Instead of forming little particles that are immediately wafted away from the flames, soot in space resembles long, curly, black, spaghetti-like strings. Materials in space burn differently than they do on Earth. In space, paper requires twice as much oxygen to burn as it does on Earth. Plastic burning in space spits out tiny fireballs in all directions.

Fire control begins with a temporary shutdown of all ventilation systems so the fire won't spread easily. Medical kits and fire extinguishers are located within easy grasp throughout the *ISS* as well as *Terra Nova* and the *Atlantis* surface habitat.

WASTE MANAGEMENT

There is no trash pickup in space. While communities on Earth are becoming increasingly concerned about the waste they generate, the tiny, Mars-bound community of astronauts has to be aware of every particle of garbage that accumulates. Most biological waste is recycled. As unpleasant as it may seem, water in urine is recovered, purified, and reused. Ammonia and nitrogen are extracted from excrement and made into fertilizer for the on-board garden. In the event of a breakdown in the waste disposal/recycling system, waste contamination will become a major health threat. Under normal conditions, the trash generation rate is 7.2 cubic feet (2.19 cubic meters) per day for a crew of 6, based on *Mir* and *ISS* studies. A journey to Mars, estimated to take 582 days, will therefore generate 4,190 cubic feet (1.27 cubic kilometers) of trash.

There have been instances in which trash has been deliberately ejected into space: Worn-out pieces of the *Hubble Space*

Telescope were chucked overboard with the knowledge that they would burn up in Earth's atmosphere. But such instances are rare. While there is a haze of debris forming a ring around Earth made up of paint chips, fragments of dead satellites, boosters, and so on, this debris has accumulated by accident. Any large object allowed to float aimlessly in space is a significant hazard to future missions and no one creates such a hazard on purpose. Essentially, when it comes to garbage in space, the three Rs (reduce, re-use, recycle) are adhered to. Whatever is left over stays on the spaceship.

The chart below gives an indication of the varieties of waste that have to be accounted for.

NON-RECYCLABLE WASTE PRODUCTS

Batteries	All types of batteries (e.g. Ni-Cad, alkaline)
Biological/Biomedical	Any solid or liquid that may present a threat of infection to humans, including nonliquid tissue, body parts, blood, blood products, body fluids, and laboratory wastes that contain human disease-causing agents. Also, used absorbent material saturated with blood, blood products, body fluids, excretions, or secretions contaminated with visible blood or blood products that have dried.
Sharps	Needles, syringes, or any intact or broken objects capable of puncturing, lacerating, or otherwise penetrating the skin (e.g., glass, scalpels, broken plastic, syringes).
Chemical hazardous	Any solid, liquid, or semi-solid trash contaminated with a chemical substance that requires special handling during disposal.
Radioactive	Solid, liquid, or gaseous materials that are radioactive or become radioactive and for which there is no further use.
Normal refuse	Any and all trash that does not meet any of the definitions above. For ground handling, normal refuse is material that has been determined to be trash that does not meet any definitions and/or criteria for regulated wastes under any federal, state, or local agency.

CLOSE ENCOUNTERS

The ever-present risks, ranging from mental and mechanical breakdown to sickness or even death, will be a constant preoccupation of Mars-bound space travelers. The voyage is a long one. Chances are they will not only think about but also have to respond to some emergencies along the way.

But the worry is bound to be offset by the sheer adventure of the voyage. Every morning the six astronauts on board *Terra Nova* feel the adrenaline rush that comes from waking up knowing that they are thousands of miles away from Earth. The view from the portholes is an incredible kaleidoscope of shifting stars and planets. No one has done what they are doing. They are, in the truest sense of the word, pioneers.

After *Terra Nova* clears the orbit of the Moon, the spacecraft begins its spin-up to induce artificial gravity. The spin-up is slow, to allow for a gentle introduction to a 1.0-G gravity environment. The vehicle is programmed to spin down gradually when the spacecraft gets close to Mars to match the 0.7-G Martian environment.

Terra Nova follows a path that takes her crew close to the planet Venus. This trajectory is designed so that the vehicle cuts across the inner solar system to reach Mars a few months before Mars arrives at opposition. There was some controversy in the aerospace community over the selection of this flight path. It brings space travelers closer to the Sun and thereby increases the risk of radiation exposure. However, the consensus among the engineers and the astronaut corps was that the crew hab is sufficiently shielded to protect the astronauts, and the Venus flyby trajectory was chosen. Flying past Venus also provides a gravity boost (see next page), and it affords humanity's first close-up glimpse of the second planet from the Sun.

July 4, 2030. *Terra Nova* approaches to within 3,200 miles (5,100 kilometers) of Venus. The crew spends much of the encounter on the observation deck, entranced by the extraordinary scale and violence of the lightning storms flickering through the Venusian clouds.

During the flyby, corrosion is found in some of the contacts between the cooling lines in different modules of *Terra Nova*, the result of two dissimilar metals coming in contact with each other. When humidity levels are high, different metals can react corrosively at their points of contact. It is a small problem, but it requires immediate attention.

2030:07:04

VENUS FLYBY

The *Terra Nova*'s flight plan is an opposition class trajectory, which will initially take the spacecraft toward Venus for a gravity assist before heading to Mars. The advantage of this trajectory is the gravity boost that a flyby of Venus will provide. As a spacecraft (or a comet or meteor) approaches a planet, it gets pulled by that planet's gravitational field. Gravity causes the spacecraft to accelerate, giving it a free energy boost, like a car rolling down a hill. As long as the spacecraft doesn't get too close, it flies away faster than when it approached. The tradeoff is a longer flight in exchange for a reduced quantity of fuel that the spacecraft must carry. The technique is common with robotic probes bound for the outer planets like the *Galileo* probe to Jupiter, and the *Cassini* probe to Saturn, but it has never before been done with a manned spacecraft.

Terra Nova's outbound flight with a gravity boost from Venus will take 11 months. The return flight, occurring while Mars is at opposition, (closest approach to the Earth) with no other gravity boosts, will take 6 months.

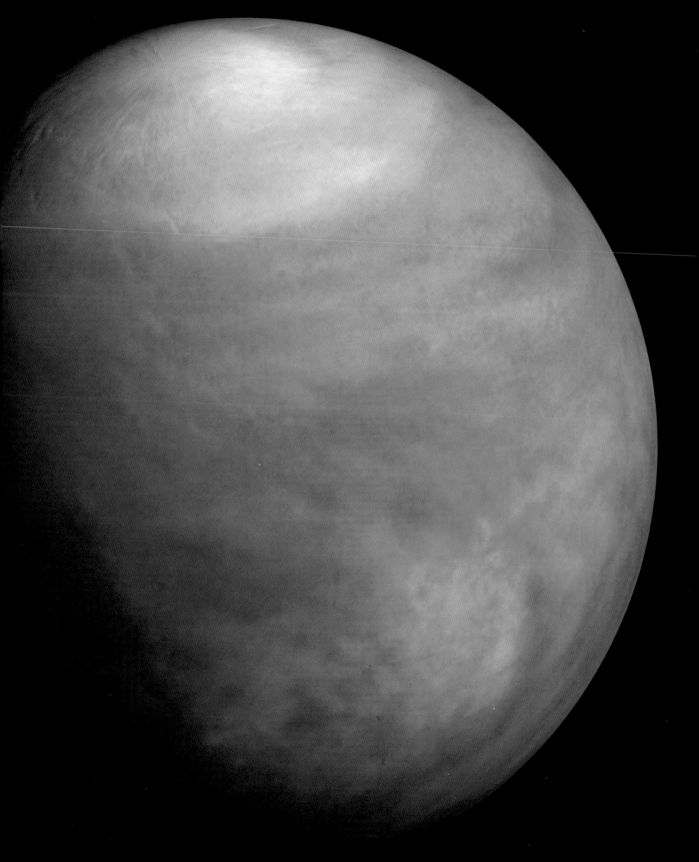

July 20, 2030. A previously uncharted near-Earth asteroid (NEA) is observed passing *Terra Nova* at a distance of a mere 9 miles (15 kilometers), a frightening near-miss. The asteroid is slightly more than 1 mile (2 kilometers) across and almost ⅔ mile (1 kilometer) wide. It has a slow, lazy spin, which affords the crew an unanticipated close-up look. As of the writing of this book, only two NEAs have been visited by spacecraft: 25143 Itokawa, by JAXA's *Hayabusa* mission; and 433 Eros, by the NASA/Applied Physics Laboratory's near-Earth asteroid rendezvous (NEAR) probe, which successfully made a soft landing on the asteroid. By 2007, only 1,000 NEAs had been identified. The largest is the 20-mile (32-kilometer) 1036 Ganymede. As of 2030, the exact number is still not known, but it is estimated to range in the tens of thousands. Some estimates place the number of NEAs larger than ⅔ mile (1 kilometer) in diameter at about 2,000. A NEA like the one that passes close to *Terra Nova*

wiped out the dinosaurs, an event known to paleontologists as KT. (The letters stand for the Cretaceous-Tertiary boundary in the fossil record: Below the KT mark, in the Cretaceous period, dinosaur fossils are plentiful, but above the KT mark, in the Tertiary period, dinosaur fossils are absent.) If the object that *Terra Nova* observed ever struck Earth, it would bring with it a level of destruction similar to the KT event.

On February 12, 2001, the NEAR spacecraft made a soft landing on the asteroid 433 Eros, pictured below. The space between Venus and Mars may be filled with hundreds of sooty, uncharted asteroids like 433 Eros.

All systems are functioning as they should. *Terra Nova* is hurtling through space at a speed of roughly 25,000 miles per hour (40,000 kilometers per hour). The waning sunlight glints off its ceramic, plastic, Mylar-wrapped form. Inside, the machinery hums beneath strains of pop and classical music as the crew goes about its tasks. Even on such an unprecedented journey, life settles into routines. The crew takes part in numerous taped interviews with a variety of media, schools, and community groups. Many hours are spent running on the treadmill and using other exercise equipment; playing chess and other board games; reading the latest research in the astronauts' respective fields; and studying charts, satellite photos, and accurate, three-dimensional representations of the landing zone. That faraway patch of real estate they're bound for is now as familiar to them as their own backyards. The crew also does some stargazing. The view outside the portholes is a constant reminder of the extraordinary, existential nature of their mission. All patiently await the next milestone in their long, long journey.

Mars lies dead ahead.

ENCOUNTERS

CHAPTER FIVE

COME IN, MARS

December 23, 2030. A single announcement says it all: "HUMANS ON MARS!" Everywhere there are celebrations and fireworks. Virtually all other human activities grind to a halt as people around the world marvel at the transmissions from Mars. A live videostream shows astronauts — not robots — preparing to step off the *Gagarin* lander's egress platform.

For a few moments, humankind puts aside its differences and holds its breath. One unified world collectively experiences the opening of a new frontier on another planet. For some, these few precious moments of global unity alone justify the cost of the expedition.

The mission commander gets his first close-up look at Mars from *Gagarin*'s egress porch.

The rover *Spirit* recorded this image from inside Gusev Crater on February 19, 2006. The slope in the foreground, topped by a house-sized plateau, was most likely formed by some sort of violent, catastrophic event, such as the explosion of a volcano or a large meteorite impact.

When, a few hours earlier, the mission commander announced that they had landed, it marked the first time that a species from Earth had become multiplanetary, microbes on meteors notwithstanding. If humans establish themselves in a permanent colony, this first landfall will be as historically important as the Cambrian explosion or the day animals emerged from the sea.

On *Gagarin*, preparations are made for the first steps on Mars. This will be one of the most highly choreographed events in the history of space exploration. The multinational nature of the first Mars landing presents thorny political challenges that *Apollo* mission planners didn't face because *Apollo* was strictly an American venture.

The first words from Mars require tough decisions. "I claim this land in the name of the Queen" is not going to fly. If everyone speaks at the same time, the first words from Mars will be garbled. Garble, by the way, was officially blamed for Neil Armstrong flubbing his famous line. What he meant to say was "That's one small step for a man, one giant leap for mankind," thus linking the actions of an individual with hope for the rest of humanity. In the excitement of the moment, however, the indefinite article "a" was apparently left out. What he said was, "That's one small step for man…." It changed the meaning of the sentence

LANDING

These extra-vehicular-activity (EVA) suits are assembled from modular units based on the specific needs of the mission. For walking on Mars, the suit has a dust visor and anti-static electrodes that keep toxic dust from working its way into the fabric. For walking in space the suits have mini jet-packs.

completely. Recent analysis of the audio has suggested that Armstrong actually did recite his line correctly and that static was to blame for the breakdown in communication.

Although English is the official language of international civil and military aviation, the first words from Mars reflect the international scope of the mission and embrace Earth's cultural diversity. The line chosen conveys the genuine idealism and hope for the future shared by the planners and the astronauts themselves. It is heard perhaps by the biggest audience ever: an astounding moment in Earth's history.

In this single moment, the astronauts on Mars' surface come to terms with the immensity of their achievement, the scale of the victory that has been won. It is a victory on many levels: They have overcome complex engineering problems as well as personal, physical, cultural, and psychological challenges. After months of being packed like sardines, they can now stretch and enjoy the sensation of real gravity. Unlike their *Apollo*-era predecessors, however, the crew of *Gagarin* do this alone. By the time their Earth-bound compatriots see the first footfall and hear the first words from Mars, the event has slipped away because of the time lag in transmission between the two planets.

When the arrival ceremony is over, the astronauts get on with the job they have been sent to do.

Dao Valles, where *Gagarin* lands, consists of two canyons that merge into a single valley, roughly 750 miles (1,200 kilometers) long and 25 miles (40 kilometers) wide.

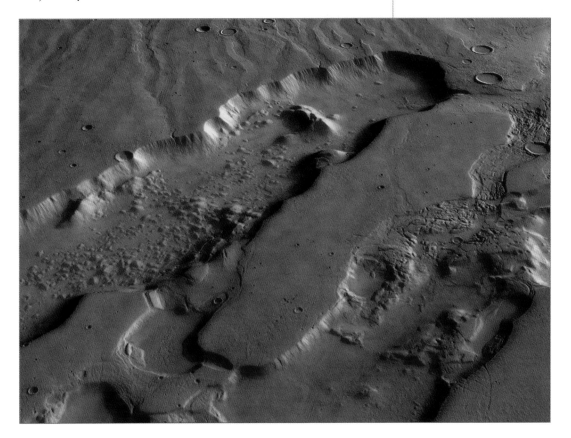

THE FIRST DAY

Now that they have finally arrived at their destination, the crew members have a lot to keep them occupied. And, no doubt, they are anxious to get started. Part of the team begins gathering contingency samples — rocks and soil — in case the mission suddenly has to be aborted and they have to return to *Terra Nova*. The remaining astronauts prepare to travel to the Mars *habitat*. A Martian dust storm forced an early landing, and now that storm is approaching fast. Another, smaller disturbance also is moving toward them.

Even though they know what's coming, the crew is thrilled by the sight of the smaller cloud of dust bearing down upon them. It is caused by the two six-wheeled surface exploration vehicles (SEVs) that have responded to activation and homing signals from *Gagarin* and now are rolling their way. They are similar to the *Apollo* lunar rovers but are larger and more powerful, capable of carrying three astronauts each as opposed to *Apollo*'s two. Like the *Apollo* rovers, they are unpressurized, open-air electric cars. Part of the crew boards one of the SEVs and sets off for the *Atlantis* Mars habitat about 1 mile (2 kilometers) away.

The *Atlantis* has more than 12,800 cubic feet (1,200 cubic meters) of living and working space and sits about 3 feet (1 meter) off the ground to allow access underneath. It's made of a double-hulled titanium-polyethylene laminated skin suspended inside an aluminum-titanium octagonal frame. The landing gear and descent motors are attached to this frame, as is the egress porch on the far end. The module is lined up north-south in an orientation that provides the maximum sunlight-filled southern exposure for the module's long axis. This orientation helps keep the module warm. The astronauts joke that it also helps with its feng shui.

From the egress porch the astronauts have to pass through the airlock before they can enter the living space. This airlock is divided into two chambers. Both chambers are designed not only to provide a pressurization transition between interior and exterior but also to form the first line of defense against dust.

THINGS TO SEE AND DO

In anticipation of the approaching dust storm some of the astronauts deploy a thousand tiny, self-inflating Mylar balloons in an experiment dubbed Balloon Atmospheric Radar Survey Over Mars (BARSOOM). Conceptually this experiment is similar to one seen in the film *Twister*, in which meteorologists track tornadoes in the American Midwest. As in the movie, the BARSOOM balloons are swept into the approaching maelstrom and tracked by radar: the motion of each balloon is then charted by computer.

The spacious airlock on *Terra Nova's* trans hab module is designed to withstand fast temperature and pressurization changes. The airlock on *Atlantis* has to guard as well against toxic dust and grit from the Martian terrain.

The caterpillar-like form of *Atlantis* is anchored to the ground by two caissons underneath the spacecraft. However, one of the landing pads has not fully deployed, threatening the craft's stability in a storm.

What emerges is a three-dimensional dynamic model of a Martian dust storm. In addition to providing information about altitude and wind speed, each balloon contains a wafer-thin transmitter that sends out information about changes in temperature. When large dust storms become global, they can temporarily raise the temperature by as much as 30 degrees.

The storm season on Mars starts some time after the planet makes its closest approach to the Sun (this is called the *perihelion*). The season doesn't normally begin until 1 or 2 months after

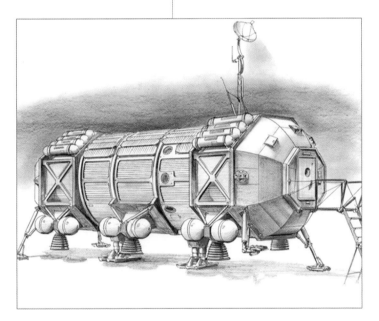

perihelion in the latter half of the Martian summer. In this particular case, in early 2030, it appears that the storm has originated in the Hellas Basin, a gigantic 3,000-foot- (9-kilometer-) deep crater in the southern hemisphere, and faltered before it spilled from the basin.

Once the BARSOOM balloons are deployed, all the astronauts, including those who were gathering samples, retire to the safety of the *Atlantis* module. They can only hope that the storm is short-lived. On Mars, big dust storms can last weeks or even months. They often become global because there are no oceans to stop their progress as there are on Earth. Sunlight is blocked, visibility is reduced, and outside the hab it is generally gloomy. Wind speeds vary from 60 miles (95 kilometers) per hour to gusts of about 100 miles (160 kilometers) per hour, but because the atmospheric pressure is so low, only the strongest gusts are felt by the astronauts. What the storm lacks in physical push, however, it makes up for by scouring *Atlantis*'s outer shell with fine grit. The sound of the sand striking the hull reminds one astronaut of summer rain tapping on an aluminum awning. Lightning discharges flash everywhere, and the sound of Martian thunder is like distant machine-gun fire. The thunderclaps are so prevalent that scientists have speculated that lightning, in addition to the ultraviolet (UV) rays of the Sun, is the cause of the toxicity of Martian soil. During the storm, hydrogen peroxide snow falls amid the sandy swirl. It is definitely a time to stay indoors.

DUST-OFF

Although there has been no exhaustive survey of the distribution and makeup of toxic chemicals in the Martian *regolith*, or soil, it is known that the soil itself and airborne dust from this soil have been highly oxidized. This oxidation is the result of billions of years of unfiltered UV rays bombarding the Martian surface. The regolith is comparable to powdered bleach ($NaBO_3$, sodium perborate) or lye (NaOH, sodium hydroxide). Since these substances are harmful to life on Earth, they may have sterilized Mars as well. And the soil almost certainly will be detrimental to the health of astronauts. It is so toxic, in fact, that it may turn out to be the one insurmountable obstacle to near-term human occupation.

Martian soil, in the form of dust, is not the same as the dust we encounter in our homes and surroundings on Earth. That's largely dead skin cells, pet dander, and mites — soft, fluffy, and icky. The dust on Mars is better thought of as grit, not dust, and it's a problem. That grit gets into every corner and crevice, every

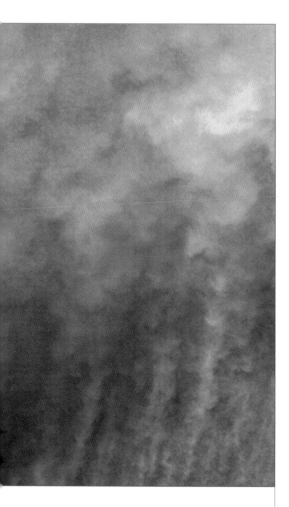

This *Mars Global Surveyor* aerial view of a Martian dust storm close to the Martian South Pole is roughly 2 miles (3 kilometers) wide and is similar to the storm that blankets Dao Valles after the crew arrives.

joint, and every pore in the skin. It is sharp and abrasive as well. The closest thing we have to it here on Earth is the fine, powdery ash that comes from volcanic eruptions. When Mount St. Helens, the volcano in Washington State, erupted in 1980, the dust was found inside car engines as far away as New York City.

When *Apollo 17* astronaut Jack Schmitt developed the runny nose and watery eyes of an allergic reaction, lunar grit turned out to be the culprit. Even though he was sealed in a space suit when he wasn't inside the lunar lander, the grit still got into Schmitt's nasal passages. Somehow it was transferred from the space suits into the lander. The good news for Schmitt was that his exposure was brief. On Mars, inhalation of this acidic grit could result in lung injury in the form of silicosis, or it could cause other forms of organ damage. Some of the components of Martian dust are known carcinogens: Cancer may be another long-term consequence of exposure.

Before the astronauts enter *Atlantis*, they use brushes and an air hose to blow off excess dust from their space suits. The suits themselves have a system of oscillating electric fields in the fabric that cause some of the dust literally to hop off. Inside the airlock, negative air pressure removes yet more dust. On Mars, the dust has a large concentration of iron, which is what gives the planet its rusty hue. Magnets located around the threshold and walls of the airlock further reduce the amount of dust and grit that gets inside. It has also been suggested that low-intensity microwaves may cause the iron-laden dust to fuse and clump together until it falls off the suit.

AFTER THE STORM

Eventually the winds die down, the peroxide snow and dust settle, and the storm comes to an end. The crew emerges to find sand drifts around the SEVs and other equipment. The next few hours are spent carefully removing the dust so that it doesn't ruin the equipment.

LEFT: After the storm, several EVAs are conducted to assess the damage to *Atlantis* and its equipment. Three astronauts climb onto one of the rovers to inspect *Gagarin*.

ABOVE: It is nightfall by the time the EVA crew arrives at the BARSOOM experiment to inspect the damage.

The next order of business is to set up the drill rig and begin the search for both water and subterranean microbes. Core samples from the drill will be cataloged and packed for analysis on Earth. Random samples of the core are chosen and sifted for microbes by the biologist in the field lab. The site for the drill is almost 1 mile (about 1.5 kilometers) away, near the Dao Valles canyon wall, where the topology suggests the kind of erosion made by melting permafrost. On Earth, a cliff formed from thawing permafrost can be found along the shores of Lake Taimyr in the heart of the Taimyr Peninsula in Siberia. What is striking about Taimyr is that the cliff face and the adjacent beaches are littered with Pleistocene fossils. On Mars, the astronauts hunt for fossil relics of microorganisms as well as living microbes. The cliff face at Dao Valles may not have woolly mammoth tusks protruding from the surface, as there are in Siberia, but there may be fossilized microbes dating from long ago.

The geological variety of Dao Valles is so rich that, in addition to the permafrost there, the prospect of finding glacial ice also makes it a likely place for water. Martian morphological features suggest that Mars went through a glacial period as recently as 5 million years ago. The causes of such glaciation are just as mysterious here on Earth as they are on Mars.

A number of construction and engineering experiments are conducted as well. In one series of experiments, the astronauts test a small furnace designed to make fiberglass sheets, pellets, pipes, and bottles out of Martian sand and air. In another, they attempt to manufacture bricks and test them for durability and ease of manufacture. This will help the second and third expeditions to Dao Valles build the infrastructure needed for a permanent base.

A day on Mars is roughly
39 minutes longer than
a day on Earth. The
resemblance is so close
that engineers have
invented a 24-hour clock
for the astronauts on
Mars in which the hours,
minutes, and seconds are
"stretched" to conform
to Martian conditions.

Not far away is the Hadriaca Patera volcano, which is thought to have old lava tubes. These are formed when the surface of a lava stream cools and solidifies while the still-molten interior continues to flow and drain away, creating an underground tunnel. Barker's Cave in Australia and the Thurston Lava Tube in Hawaii are the largest lava tubes known on Earth. Martian lava tubes are likely to be immense caverns much larger than those on Earth because Earth's stronger gravity quickly pulls down any large tubes. Even on Mars the tubes have likely collapsed at one or both ends, just as they do on Earth. The air, water, and soil that entered the lava tubes before the collapse is consequently sealed off from the rest of the harsh Martian climate.

Such entombed spaces might in effect constitute a time capsule, affording scientists a direct sample of an earlier epoch in Martian history. These sealed chambers could also be micro-

All six astronauts enjoy their sensational emergence from the somewhat claustrophobic quarters in which they have been confined to the wide-open, if desolate, Martian landscape.

habitats to relic species of chemophyllic micro-organisms unique to each sealed chamber. Such a finding could, in our most daring dreams, reveal a Martian version of the Galapagos Islands, with distinct communities of microbes dwelling in each chamber, sealed off from one another, and from the rest of the Martian climate, each making its own way down a distinct evolutionary path or perhaps not evolving at all but retaining its original form. Similarly, the discovery of water — where it might have been geothermally heated by a nearby volcanic vent — is as likely in a lava tube as anywhere.

Many scientists believe that Hadriaca Patera had significant interaction with subsurface water, producing mostly explosive pyroclastic flows of ash rather than just lava flows. Dao Valles, on the southern slopes of the volcano, may have been the source of this water.

THE PERILS OF MARS

The very thing that excites scientists about going to Mars — the prospect of finding microbial life — may also pose the greatest danger. An organism that can survive the toxicity of Martian soil, the extreme temperatures of the Martian climate, and daily bombardment by unfiltered solar radiation could become an exceedingly robust pathogen for which there may be no known antidote. Without knowing what we might be up against, it is virtually impossible to have vaccines or treatments ready ahead of time. An astronaut who contracts a Martian pathogen might have to be quarantined for the rest of his or her natural life. But, in a 1997 study, the National Research Council came to a more hopeful conclusion. "The chances that invasive properties would have evolved in putative Martian microbes in the absence of evolutionary selection pressure for such properties is vanishingly small. Subcellular disease agents, such as viruses and prions, are biologically part of their host organisms, and an extraterrestrial source of such agents is extremely unlikely." In other words, in all likelihood, no such pathogens exist.

Like the dangers posed by the chemical toxicity of the soil, the threat posed by Martian microbes is as yet unquantifiable. Not knowing where or if a danger exists means that every possible contingency will have to be covered. As we become more familiar with the Martian environment and the specific dangers it poses, we'll be able to tailor our countermeasures accordingly, thereby reducing the costs. One proposal suggests that a site to be explored be pre-qualified: A robot could be sent ahead to ensure that the local toxicity is low. Only when the area has been declared safe would humans decide how to proceed. What makes this "zoning" beneficial is that such a robot reconnaissance could also check for signs of biological activity.

If the dangers we don't know about are frightening, those we do know about are every bit as terrifying.

A lot depends on just how healthy the astronauts are when they land. The long voyage may have weakened their muscles and bones to the point where time is needed to recover inside *Gagarin* before the first EVA is attempted. Degradation of human mobility as a result of gravity-related bone atrophy could result in serious injuries that would, at best, jeopardize the mission and, at worst, result in death. To stave off this threat, *Terra Nova's* crew has been exercising vigorously throughout the voyage.

The low atmospheric pressure on Mars poses a constant threat to the lives of the astronauts. A breach in a space suit or in *Atlantis's* hull could become a slow leak or, if the breach was large and sudden, lead to instant depressurization. A slow leak could be

DANGERS

difficult to find and waste precious air, but a breach that causes instant depressurization would be catastrophic. The physical consequences are horrible to contemplate: Nitrogen gas in a person's bloodstream would fizz as if escaping from an open soda can, causing blood to ooze from every opening in the body. Retinal hemorrhaging and cerebral edema would immediately commence, causing blindness, hallucinations, bizarre behavior, loss of coordination, paralysis, and coma — conditions consistent with high-altitude cerebral edema and the bends. Death would follow a minute or two of excruciating agony. Definitely not a pleasant way to go.

Findings from the rovers *Opportunity* and *Spirit* have suggested that, at any given moment, nearly half a dozen dust devils can be seen along the Martian horizon. Dust devils on Earth are mild eddies compared to their Martian cousins. The scale and the speed of dust devils on Mars are more akin to those of terrestrial tornadoes, but they don't have a lot of force behind them because of the low atmospheric pressure. The danger posed by a Martian dust devil or dust storm stems from the static electricity it generates, not from an actual tornado-class wind. This makes the chances of getting zapped by an electrostatic discharge from the equipment, the SEV, or the hab itself relatively high. Local discharges and arcing will be a constant hazard.

Then there's radiation. Again, because the atmosphere on Mars is thin, it offers little protection from the Sun's rays. On Earth, our atmosphere filters out about 95 percent of the Sun's rays. X-ray, gamma-ray, ultraviolet, microwave, and some infrared radiation from the Sun are all blocked. Only visible light and radio waves can get through our atmosphere. But that's not true on Mars. All of it reaches the ground. Radiation shielding is a must, not only for the flight between the planets but for the stay on Mars as well. Eventually astronauts will want to build infrastructure underground or select sites that are at least partially in the shade.

THE WORKING DAY

Every day on Mars, there are at least four people on an EVA, that is, in a space suit and outside the hab. Sometimes, all six may be outdoors. The buddy system is in effect at all times: No one can be outside alone. At least one member of the crew is always at hand to assist another.

Topping the list of chores is daily work at the drill site. Core samples are pulled off in 6-foot (2-meter) lengths, tagged, sealed in plastic tubing, entered into the database with notes and digital photos, and then placed in a storage bin for transport back to *Gagarin*. The drill is kept running only during daylight hours, although there is a capacity for working at night as well. While it is in operation, one person monitors telemetry from the drill, that is, its angle of entry, the depth attained, the temperature behind the bit, the consistency of the rock, the rock's chemical content, and (wishfully, perhaps) the presence or absence of water. All of this data is sent to the computer and is recorded with the core sample. A second person watches the drill's vital signs, making sure that it is running smoothly and not overheating or working beyond its capacity. Should something go wrong with the drill, the astronauts are well trained to improvise a solution. No matter how mundane the task, the crew is always aware of the possibility of an amazing breakthrough of some kind, especially the positive discovery of water or proof of life.

A host of other experiments need tending to as well. A typical day on Mars may include time monitoring the drill, collecting data, or exploring the surrounding terrain. Training

Heave, Ho! The commander and two of the crew apply old-fashioned, low-tech brute force to the balky landing gear.

at Mars analog field stations on Earth has given the astronauts a sense of what the routine will be like, but nothing can prepare them for the reality. Think of it! Their feet are the first ever to leave traces on Mars!

But getting around on foot is not as easy as one might think. There is a widespread misconception that the lower gravitational pull makes it easy to bound like a gazelle across the landscape. In fact, the weaker gravity makes it harder. As the pull of gravity falls off, it gets progressively more difficult to move fast. Both Moon and Mars walkers miss the pull of gravity that draws their limbs downward and torques the body's center of mass forward. It is difficult to move at any decent forward speed with a normal stride. For this reason, the *Apollo* astronauts took to hopping around. Even when they are not in a bulky EVA pressure suit, it's the easiest way to move forward.

Everyone wants desperately to make every minute of every Martian day count, and that pressure comes as much from the astronauts as it does from Mission Control. One of the astronauts says that working on Mars is like running through a grocery store with a shopping cart a few minutes before the store closes, trying to grab as many items as possible from the shelves. They know that there will be ample time to rest on the way home, and this knowledge helps to keep them going.

When the *Apollo* astronauts were on the Moon, the fieldwork they conducted was heavily scripted. At the start of their EVA, they trekked as far away from the lunar lander as they could and slowly worked their way back, examining pre-selected sites along the way. On Mars, the style of fieldwork will more closely resemble fieldwork on Earth. The astronauts will, of course, have pre-defined tasks, but they also can explore points of interest in a somewhat ad hoc, unscripted manner.

FIELDWORK

This reactive approach is perhaps the most compelling technical argument in favor of sending humans to Mars. Robotic missions can't respond as well to their discoveries. Humans can make real-time decisions: to conduct experiments, process information, and then create new experiments on the fly. This approach is good for science, exploration, and research. It also allows astronauts greater flexibility to deal with contingencies or emergencies.

Mars explorers also will benefit from the significantly longer time they'll spend on Mars. By comparison, the *Apollo* astronauts' longest stay on the Moon was the four days of *Apollo 17*. The *Terra Nova* crew will spend 60 days on Martian ground.

CASA DAO VALLES

To ask one of the crew members domiciled in the *Atlantis* what living on Mars is like is to invite any number of comparisons to life on Earth. One of them jokes about writing an article for *Trailer Life*, a magazine devoted to open-road camping. His article would be entitled "My Winnebago on Mars" and would list the many advantages Mars has over Earth for the cross-country camper. There are no bears rummaging through trash cans at night, for example, and no park rangers imposing fines for excessively boisterous behavior. Campfires on Mars, however, are out of the question.

All joking aside, the astronauts find that *Atlantis* is small but comfortable. It boasts some snazzy, innovative details to ensure that space is used as efficiently as possible, all the product of years of careful study of the ergonomics of living within a sealed environment. Since the astronauts will be arriving in their new home during the winter holidays, the ground crew back on Earth has

"They had a house of crystal pillars on the planet Mars by the edge of an empty sea...."

— RAY BRADBURY, *THE MARTIAN CHRONICLES*

secretly decorated the hab as a nice surprise. Little things like this can have an enormous, positive psychological effect.

In keeping with spaceflight tradition, each of the 60 days on Mars begins with a musical wake-up call, usually some Mars-themed music ranging from Gustav Holst's "Mars, the Bringer of War" from *The Planets* to David Bowie's "Spiders from Mars." On several different occasions, political leaders from participating nations have phoned in a wake-up call, but those were highly choreographed events that allowed little in the way of spontaneous fun.

In the early days of the space race, astronauts ate cubes of gelatin-coated sandwiches and a kind of nutritious but unappetizing paste squeezed out of tubes like toothpaste. By the time of the *Apollo* missions, meals had a more familiar look and taste. Today's interplanetary explorers are served food that differs only

Even life on Mars has its routines. The doctor files a daily medical report for each member of the crew.

marginally from the kind of lightweight items that campers take on backcountry expeditions.

Breakfast on Mars begins before sun-up and consists of tea or coffee brewed from sealed packets and reconstituted hot or cold cereal and fruit. In addition to packaged and freeze-dried food, the crew enjoys fresh vegetables and produce grown in the hydroponic garden units brought down from *Terra Nova*. Lettuce, radishes, carrots, and wheat germ provide fresh supplements to the daily menu. Meat products also are reconstituted from freeze-dried foil packets. NASA has looked into the possibility of growing fish meat — not whole fish, but rather just the muscle tissue. In one experiment, cubes of fish flesh were washed in alcohol and placed in a vat of bovine serum, a nutrient derived from calves' blood. After a week, the fish chunks had grown by 16 percent. This approach would allow fresh meat to be raised without the mess of slaughter. Mission planners disallowed this potentially useful, alternative protein source, however, to enforce a general prohibition on living flesh products in *Terra Nova*'s pantry — a safety concern.

Food preparation techniques both on Mars and in space are not quite the same as in a kitchen on Earth. All kitchen utensils on *Terra Nova* and in the space hab *Atlantis* are made of super-strong, lightweight zirconium oxide ceramic. Waste disposal and cleanup using pre-moistened napkins are attended to immediately. In the compact galleys of homes in space, odors can very quickly become overpowering.

Breakfast is followed by domestic chores. Each day, two people take a turn at washing their clothes, which means that each individual's clothes are washed every 3 days. The water is completely recovered from the wash and analyzed before it is recycled. The analysis provides another way to check whether or not Martian dust has infiltrated the cabin.

As the *sols*, or Martian days, pass, each one crammed with mundane chores and sometimes fascinating experiments, the astronauts can never forget that they are inhabiting a dangerous, unforgiving environment. They can never act carelessly, especially outside the hab, where even a small misstep could be fatal. But humans have a wonderful capacity to adapt to difficult conditions and the astronauts are especially well-suited, both by character and training, to take risk in stride. Chances are that they will savor every moment on Mars. Their 2-month sojourn will give them ample opportunity for both practical experimentation and philosophical thought. There is also a real possibility that they will change forever the way we look at the Universe. They might just discover evidence of Martian life.

The exotic landforms of the northwest Hellas Planitia are a cornucopia of oddly eroded surfaces seldom seen from space. This 2-mile- (3-kilometer-) wide region near 39.3°S, 306.7°W is usually clouded over but a break in the clouds in October 2003 allowed the *Mars Global Surveyor* to capture the picture.

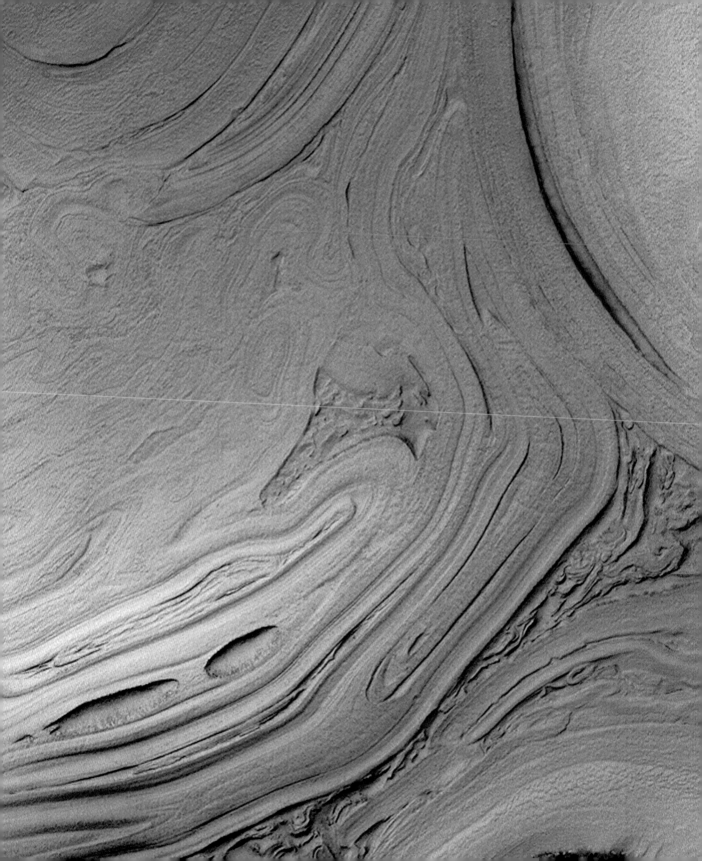

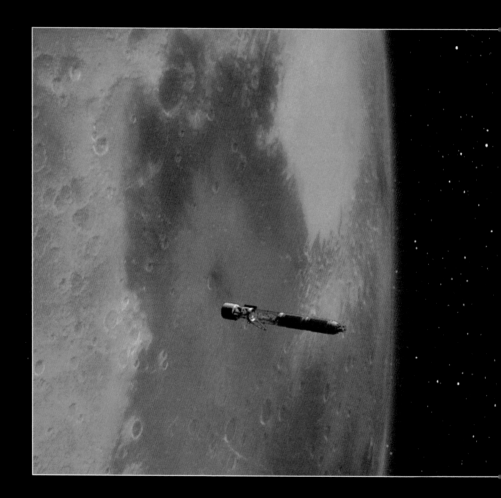

LIFE
ON MARS

> "FAR-STRETCHING, ENDLESS TIME
> BRINGS FORTH ALL HIDDEN THINGS"
> — Sophocles, *Siege of Troy*

Mars and the question of life are two inseparable topics, thanks in large measure to Percival Lowell and his dying Martian canal builders. No one can speak of life in space without speaking of Mars, and no one can speak of Mars without pondering life.

We know of one genesis event in the history of Earth. Could there have been a second genesis on Mars or elsewhere in our solar system?

The now-vacant *Terra Nova* awaits the return of the astronauts on Mars.

THE GOLDILOCKS ZONE

There is a range of distances from the parent star in a solar system where water can exist in liquid form. This is called the *Goldilocks Zone* because the temperature is just right. It's not hot enough to cause water to evaporate or boil off and not so cold that water freezes. Historically, scientists believed that life could exist only in this zone, but they may have been thinking too much in terms of human experience.

Two findings in the past decade have caused scientists to look beyond the Goldilocks Zone for signs of life. The first was the discovery of a class of microbial organisms collectively known as *extremophiles* or *hypothermophiles*. These are tiny microbe communities that thrive in habitats no one would have thought were remotely hospitable: inside the ice of a polar glacier, at one extreme; inside hot lava sludge from an active volcano on Hawaii's Big Island, at the other. These organisms have been found in the toxic waste of industrial cleanup sites, and one species, *Deinococcus radiodurans*, has been nicknamed "Conan the Bacterium" because it can withstand radiation 10,000 times greater than the dose that would kill humans. The discoveries have forced scientists to consider a correspondingly wider range of habitats throughout our solar system.

The discovery of liquid water in places far beyond the Goldilocks Zone was the second finding that has reoriented the search for life. Confirmation that water once flowed on Mars has already piqued our curiosity; the discovery of water in liquid form on Jupiter's moons Europa and Callisto and on Saturn's moons Iapetus and Enceladus compels us to look for life beyond the Goldilocks Zone. Now it appears that as long as there is a source of heat, there may be oases of chemosynthetic life far away from the parent star.

"The thing that sets Mars apart is that it is the one planet that is enough like Earth that you can imagine life possibly once having taken hold there."
— STEVEN SQUYRES, CORNELL UNIVERSITY PLANETARY SCIENTIST

THE RISE OF LIFE ON EARTH

We know that life on Earth got a very early start. If it arose before the catastrophic collision associated with the formation of our Moon, life on Earth may have been extinguished. It is quite possible that life had at least one false start. William Schopf, of the University of California at Los Angeles, has found chert rocks from the Pilbara region of western Australia that have carbon isotopes and unicellular anaerobic microfossils dating back 3.5 billion years, almost to the dawn of the Archaean eon or roughly 1 billion years after the formation of Earth. Although Schopf's findings have been disputed, primarily by Martin Brasier of Oxford University, the prevailing opinion is that Schopf's fossils are evidence of the early presence of life on Earth.

What is striking about these fossils is not that they are so old but rather that they represent a form of *cyanobacteria*, a primitive bacteria sometimes misidentified as blue-green algae. These were possibly the first organisms to possess a photosynthetic metabolism; like modern plants, they need sunlight to thrive. Cyanobacteria can be found in nearly every habitat on Earth. They can be single-celled, or they can gather in colonies that form mats and strings. Remarkably, in western Australia they can still be found in small mounds called *stromatolites*. These are living fossils that date back to the Precambrian era — more than 600 million years ago — which makes them organisms far older than dinosaurs.

It is most likely that some earlier form of chemosynthetic organism appeared prior to the rise of cyanobacteria, but no fossil evidence for it has been found. There are, however, rocks that date back 3.8 billion years that have biomarkers, that is, they show chemical traces that can only be the result of biological activity.

The team drills both for water and for microbes, either of which may provide answers to age-old questions about the Red Planet.

Because the astronauts'
time on Mars is limited,
they work at the drill site
and on other experiments
in shifts that extend around
the Martian clock.

It should be noted that nothing smaller than cyanobacteria has been found in the fossil record. As far as we know, prions and viruses don't make fossils. This is unfortunate because it means that the fossil record of life on Earth has a built-in boundary beyond which we cannot see. It's like having a book that starts at the second chapter and consequently omits important information about the characters and setting.

What is missing from the fossil record of the book of life is information about the transitions between a world with no biological activity and one containing the first simple organisms.

In order to search effectively for past life on Mars, we need to understand its origins on Earth. One novel approach has been taken by Craig Venter, a pioneer in the mapping of the human genome. His Institute for Genomic Research is involved in a top-down search for the most minimal requirements of life at the genetic level. He and his colleagues are engineering prokaryotic cells with progressively fewer genomes in the hope of identifying the minimum genetic requirements for living organisms. If they succeed, they will have a blueprint to follow when they embark on a bottom-up strategy for synthesizing life from molecular compounds.

MILLER-UREY
OR HOW TO MAKE LIFE

So far, we have only a tenuous understanding of what happens in the transition between prebiotic and biological chemistry, between the non-living and life. At the University of Chicago in 1953, a classic experiment was conducted by Stanley Miller and his professor, Harold Urey. Their goal was to simulate the conditions of primordial Earth in a hermetically sealed environment and see if organic compounds could be synthesized from non-organic chemicals. Water (H_2O), methane (CH_4), ammonia (NH_3), and hydrogen (H_2) were sealed together in a set of flasks and glass tubes. The water was heated to simulate evaporation and fill the air with water vapor. A pair of electrodes was attached to the flasks to simulate lightning strikes, and the water vapor was later cooled and condensed back into liquid form. This cycle was set up to run continuously for a week.

What Miller and Urey found was that as much as 15 percent of the carbon from the methane gas was turned into basic organic compounds. Two percent of this carbon formed the kind of amino acids that make proteins (e.g., glycine) in living cells. The experiment proved that some of the organic building blocks of life could arise spontaneously in an abiotic environment, creating a rich organic soup in warm little ponds and in primordial oceans. But the experiment failed to synthesize living, self-replicating organisms and left unsettled the question of how these basic building blocks congregate into more complex forms — macromolecules like RNA and DNA that could act as precursors of living cells.

GENESIS

EXTREME LIVING

When we think of extraterrestrial life, iconic images of flying saucers and furry aliens with a penchant for human abductions spring to mind. We don't typically think of microbes as ETs, but they may be the only extraterrestrial game in town, at least in our solar system. In recent years, living microbes have been discovered in environments so harsh that they were previously thought of as uninhabitable by any form of life. But now our microbial cousins have demonstrated that our view of the Universe has been anthropomorphically distorted. In the past, we looked for life only in those habitats in which humans could exist, and we applied that same standard when we looked beyond Earth, to the solar system and cosmos at large.

Now we have a better understanding of the range of living conditions that some life forms can endure. Organisms that tolerate and thrive in habitats deadly to humans are called *extremophiles*. Their range and diversity are only now becoming known. They are of immense interest to astrobiologists, who now have reason to hope that the extreme conditions that prevail on other planets may, after all, shelter life.

This is a partial list of some of these new types of microbes — bacteria and archaea — that have been described so far:

- **Acidophiles** thrive in extremely acidic environments. They include the algae, *Cyanidium caldarium*, found in the highly acidic hot springs at Yellowstone National Park, and the archaea, *Ferroplasma acidiphilum*, which lives on iron found in industrial acid drainage in mines.

- **Barophiles and piezophiles** are bacteria that live in high-pressure environments on ocean floors, deep ocean trenches, and inside Earth's crust. *Shewanella benthica* was discovered in 1996 in ocean-floor sediments dug from the deepest ocean bottom in the world at 6.8 miles (10.9 kilometers) below the surface — the Challenger Deep portion of the Marianas Trench.

- **Halophiles and Alkaliphiles** live in environments with a very high concentration of salt (NaCl), approximately ten times the salt level of ocean water.

- **Lithoautotrophs** are capable of deriving energy from reduced minerals. In other words, organisms such as the lithoautotroph *Sulfolobus solfataricus* eat rocks.

- **Oligotrophs** function with an extremely slow metabolism. This allows them to survive in a wide range of habitats, including deep oceanic sediments, caves, glacial and polar ice, deep subsurface soil, aquifers, and ocean water.

- **Psychrophiles and cryophiles** are hardy in extremely cold temperatures. They are commonly found in permafrost, polar ice, cold ocean water, and in alpine snow pack. *Psychrobacter cryohalolentis* was found in Siberian permafrost.

- **Radioresistants** resist nuclear and UV radiation. *Deinococcus radiodurans* can withstand doses of gamma radiation 500 times more intense than that which would kill a human.

- **Thermophiles** flourish in temperatures ranging from 60° to 80°C. They are commonly found in hot springs and undersea hydrothermal vents. Thermophiles such as *Methanococcus jannaschii* interest astrobiologists because they produce methane gas, clouds of which have been found on Mars.

- **Xerophiles** live in extremely dry, desiccating conditions. Cacti are a common form of xerophilic organism. So is *Artemia salina*, a primordial crustacean better known as a sea monkey.

- **Polyextremophiles** qualify as extremophiles under more than one category. Several examples cited in this list are polyextremophilic, notably, *Deinococcus radiodurans*. It is all but guaranteed that extant microbes on Mars will be polyextremophiles.

Since the Miller-Urey experiment, several models have been put forward that are meant to describe how nature spontaneously worked its way from simple prebiotic chemistry to the far more complex composition of a living proto-organism. Each is a study in the forensics of creation.

In the 1980s, chemist Günter Wächtershäuser proposed a revolutionary new approach to investigating the origin of life, called the iron-sulfur world theory. His idea was that life arose first in hydrothermal vents — a notion that has been heavily reinforced by the discovery of entire ecosystems living around hydrothermal vents on the ocean floor.

According to this theory, a primitive metabolism in the form of a chain of cyclical chemical reactions predates the rise of genetics. The chemical reactions occur inside microcaves in the mineral walls of the hydrothermal vent. The large temperature gradient in and around the vent, combined with the bone-crushing pressure that takes effect at such extreme depths, and the flow of sulfide-bearing water through the vent allows metabolic processes to evolve in different zones around the vent. The first proto-cells to emerge from this scenario would be the oily lipid microbubbles (fat is a kind of lipid) that surround a self-sustaining metabolism.

A related theory suggests that eventually primordial seas became full of these lipid bubbles. Along the shores of these seas, bubbles collected in a frothy foam in the surf and washed onto the beach or collected in warm tidal pools like driftwood. Evaporation concentrated the bubbles, allowing the exchange of ever more complicated compounds. The beauty of this model, as Wächtershäuser has pointed out, is that the processes of creation are not limited to a single event or moment but can be continuous — they could be happening now.

NEXT PAGE: ESA's *Mars Express Orbiter* took this image of a frozen pond of water ice roughly 21 miles (35 kilometers) in diameter in the Vastitas Borealis region of Mars.

The rover *Opportunity*'s finding of a prehistoric shoreline on Mars makes this bubble-to-beach idea especially attractive. To validate it, we would need to locate a corresponding hydrothermal vent on Mars that was once under water.

When cyanobacteria first appeared 4 to 3.5 billion years ago, Earth's atmosphere was very different from today's in its chemical composition; like Mars today, the atmosphere of the Archaean Earth consisted primarily of carbon dioxide. Respirating microorganisms changed its chemical makeup over hundreds of millions of years until oxygen became a primary component about 2 billion years ago. If life arose on Mars and was similar to terrestrial micro-organisms, it may not have survived long enough to have any significant impact on the composition of Martian air. Alternatively, life on Mars may not have evolved to a state where photosynthesis was its primary means of metabolism. This may be good news for those who hope to find extant organisms on Mars today. Life that is not dependent on sunlight for sustenance has subterranean survival options not available to sunlight-loving organisms that rely on photosynthesis. Chemosynthetic organisms can live deep in the ground heedless of conditions on the surface and in the atmosphere.

Biochemist Lisa Pratt and her colleagues at Indiana University have collected microbe samples from mine shafts in Canada and South Africa. Water leaching into these mine shafts, some more than a kilometer underground, brings with it a supply of indigenous subterranean microbes. Some of these microbes live by eating rock: they literally turn stone into organic material. They are called *chemolithoautotrophs*, and the organic matter they produce is consumed by other kinds of microbes, organisms called *chemoorganotrophs*. In effect, a whole microbial community thrives without benefit of sunlight.

These communities do, however, depend on water flowing through fractures or leaching through the soil. How deep they reach is not known exactly. A borehole drilled in Gravenberg, Sweden, located thermophilic bacteria at a depth of 17,316 feet (5,278 meters). The overall size of Earth's subterranean biomass is thought to be at least equal to, if not greater than, that of all the organisms that live on Earth's surface. The prospect that Mars, too, could have a subterranean biomass may be the last great hope for life on the Red Planet today. But finding it won't be easy or cheap.

MARTIAN GEOLOGY

Some of the strongest arguments in support of a second genesis come from geology rather than biology. We can see from the geological record that Mars was once a much more active place than it is today. Its surface morphology bears this out as well, so it is fair to compare some of the striking similarities between primordial Earth and a primordial Mars.

At the beginning of the Archaean eon, Earth was nearly three times warmer than it is now. The continental drift associated with plate tectonics and volcanism was considerably greater. And Earth's crust was not only thinner than it is now but also probably broken up into many small protocontinents which combined and then broke apart to form a series of continents and super-continents.

On Mars, tectonic activity was considerably tamer but not altogether absent. Odd features, such as the Tharsis Bulge, show faulting and rifts that date as far back as 4 billion years. The super-canyon Valles Marineris is a rift valley — basically a large tear across the Martian crust — similar to the Great Rift Valley that runs along the eastern side of Africa.

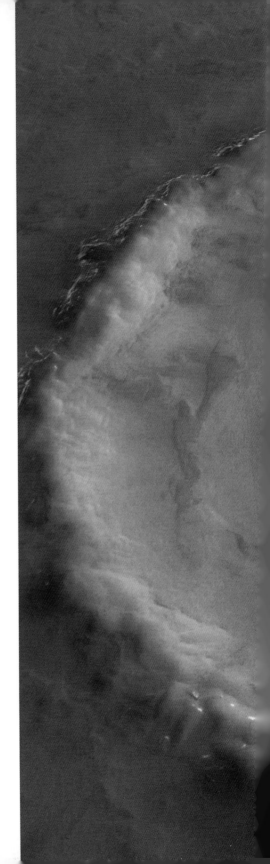

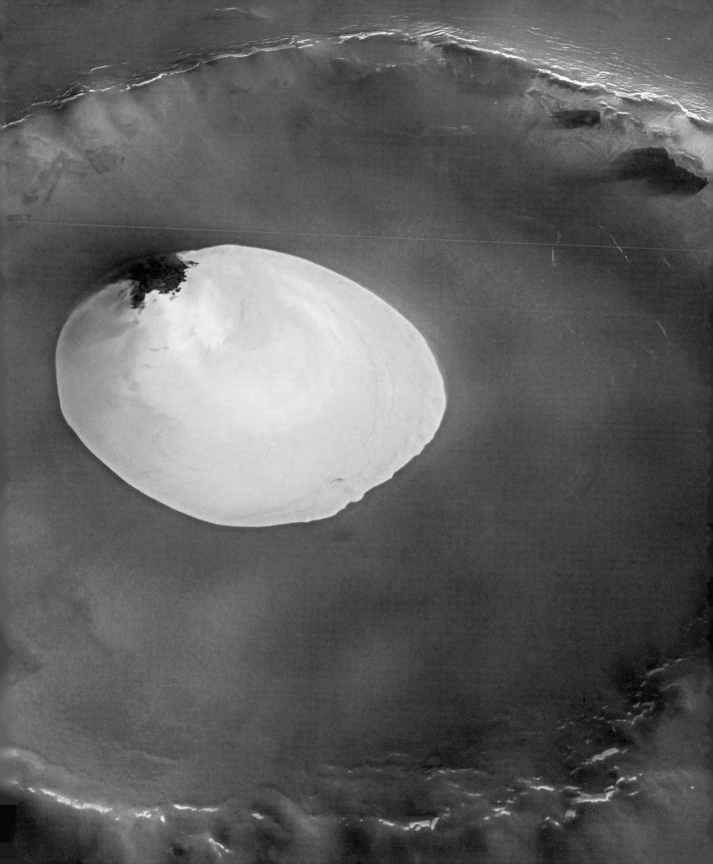

THE GREENING OF MARS

Terraforming — a process in which the climate of Mars, Venus, or some other planetary body is engineered to make an Earthlike habitat — is a relatively new feature in the history of our dreams about Mars. It was first suggested by Carl Sagan in an article written in 1961 about Venus. NASA began to examine the question when it hosted a conference on planetary transformation in 1979, and others have since speculated about a greened Mars as well, notably James Lovelock of Gaia Hypothesis fame, and NASA's Christopher McKay.

What makes Mars so attractive as a candidate for terraforming? For one thing, the length of the Martian day is almost identical to Earth's. Its 25-degree axial tilt is similar to Earth's 23.5 degrees. This tilt gives Mars four seasons, just as it does on Earth. The temperatures on Mars, although harsh by Earth standards, may be within our ability to alter.

The challenges facing those who want to terraform Mars, however, are formidable. First, the atmospheric pressure on Mars is so low that it is not even 1 percent as dense as the air on Earth at sea level. Second, the Martian atmosphere is 95 percent carbon dioxide. Some scientists suspect that the Martian polar icecaps are composed mainly of frozen carbon dioxide — dry ice. Third, Mars is so cold that only on a hot summer day at the equator does the temperature climb above the freezing point of water. Extremophiles excluded, most metabolic reactions are inhibited in such bitter cold. Certainly no multicellular organism from Earth would survive. A fourth challenge is posed by the continuous bombardment of the planet by deadly ultraviolet rays from the Sun. At ground level on Mars, the UV level is lethal, and consequently the Martian soil has been effectively sterilized. Finally, there is the apparent lack of water — without which life is unsustainable.

Many of these problems can be resolved, according to some scientists, by inducing an artificial form of global warming through the introduction of greenhouse gases. The ideal gas for UV shielding is ozone (O_3).

"Who has cut a channel for the flood water, or the path for the thunderstorm; To cause it to rain on a land where no man is; on the wilderness, in which there is no man; to satisfy the waste and desolate ground, to cause the tender grass to spring forth?"

— *JOB* 38:25–27

Alternative greenhouse gases include ammonia (NH_3), as proposed by Carl Sagan, Robert Zubrin, and Christopher McKay, and perfluorocarbons (PFCs or CF_4), as suggested by Martyn Fogg and endorsed by Zubrin. The problem of toxic soil may be addressed by making fertilizer from locally available resources. Nitrogen, for example, can be extracted from the Martian air, and potassium and phosphate exist in the ground in small but serviceable quantities. The provision of water may be addressed by further exploration. There are grounds to suppose that quantities of water are locked in a kind of ground ice in the Martian regolith. Thickening the atmosphere and warming the planet will help return that water to a liquid state.

As currently envisioned, the transformation of Mars will be achieved in two stages. The goal of the first stage, *ecopoiesis*, is to change Mars from its natural state to a condition suitable for plant life, but not necessarily suitable for human habitation. Ecopoiesis can be initiated by a variety of means. One method calls for slamming icy comets rich in ammonia into the planet. Another proposal entails thawing out the Martian permafrost with atomic waste. This would release greenhouse gases trapped in the ground. Another idea is simply to dig large, deep holes to provide a kind of vent for letting the heat from the planet's interior escape into the atmosphere.

The second stage in the transformation of Mars is terraforming proper, the process — likely to take thousands of years — by which Mars is brought to a condition that would permit humans to dwell on the surface, breathe the air, and live in a relatively stable environment.

Recently, the European Space Agency's *Mars Express Orbiter* produced a global mineral map of Mars which in turn has indicated sites where life might conceivably have existed. It shows, for example, where large bodies of standing water might have been present on Mars immediately after its formation. Using this map, Jean-Pierre Bibring of the Institut d'astrophysique spatiale in Orsay, France, has suggested that the geological history of Mars falls into three eras.

The earliest is the Phyllosian era, which occurred between 4.5 and 4.2 billion years ago, soon after the planet was formed. Mars was considerably warmer and wetter then, making possible the deposit of large clay beds, many of which survive today. The second era is the Theiikian era, which took place between 4.2 and 3.8 billion years ago. This was the age of erupting super-volcanoes that drove global climate change. In particular, sulfur from volcanic plumes reacted with water to produce acid rain, which changed the composition of the rocks and minerals on the ground. In 2004, the rover *Opportunity* detected very high concentrations of sulfur in a rock outcrop. The chemical form of this sulfur was bonded with magnesium, iron, and other sulfates, which, for some species of autotrophs, makes fine dining.

Finally, there is the Siderikian era, the longest-lasting of the Martian eras. It began from 3.8 to 3.5 billion years ago and continues today. It was and is a dry period, a time when weathering gave Mars its red color.

Some scientists speculate that of these three eras, only the first, the Phyllosian, might have supported life. The idea is that the clay beds formed at the bottom of shallow lakes and seas

Because the atmospheric pressure on Mars is so low, water gushing from a subterranean reservoir out of the drill well will immediately turn to snow and dangerous shards of ice.

might have provided the damp conditions in which the processes of life could begin. Clearly, these clay beds will be high-priority targets for future robot landers. It's possible that the cold Martian conditions have preserved some or most of the biological record in these beds.

Is Mars still geologically active? The discovery of methane on Mars is highly suggestive. On Earth, methane comes from just two sources: volcanoes and bacteria.

IT'S A GAS

Recently, the orbiting *Mars Global Surveyor* discovered major sources of methane (CH_4) over Valles Marineris and the Hellas Basin. Meanwhile, NASA's *Mars Odyssey Orbiter* detected an abundance of hydrogen, which is a sign of large deposits of water ice, in the Hellas Basin as well. Could there be some correlation between the locations of the methane and the water? These findings of both methane and hydrogen in the same locale were confirmed by Earth-based observatories in Cerro Pachon, Chile, and by the W. M. Keck Telescope atop Mauna Kea in Hawaii. Additional discoveries of methane have been made along the Martian equator.

NEXT PAGE: The rover *Spirit* at Gusev Crater in 2005, alone in the Martian wilderness.

Scientists are making a considerable effort to determine the source of these large methane clouds on Mars. On Earth, methane comes mostly from bacteria in the soil and from gas-passing animals like cows and people. Methane doesn't last long, so something on Mars is replenishing these clouds. If the source turns out to be vents and not bacteria, it will not be altogether bad news, for the vents themselves may harbor some sort of extremophile. The detection of methane, along with the clay beds, has buoyed hopes for finding organisms living on Mars today.

THE VIKING PROGRAM

Scientists, however, are cautious about making claims that signs of life have been found; they've had their hopes dashed before. When the two *Viking* landers collected soil samples with their robot scoops in 1976, they conducted three experiments to test for the presence of microbes in the soil. One experiment, called "labeled release," added nutrients to the soil, the kind of nutrients that on Earth were readily consumed and digested by microbes. These nutrients were tagged with a radioactive isotope, Carbon-14 (C-14). If there were microbes in the soil, they could be expected to release the C-14 into the air and *Viking* landers would detect a corresponding rise in the carbon level immediately above the soil sample. The results were positive: Both landers found that the levels of C-14 immediately increased, a strong indication that there were organisms living in the soil.

But then the results from a second experiment were puzzling: A gas chromatograph and mass spectrometer (GCMS) was used to search for organic compounds by heating the soil to different temperatures so that various compounds would be released and could be identified separately. It found no organic compounds.

A third experiment, the gas exchange experiment (GEX), also yielded ambiguous results. Like labeled release, it worked by sampling the gases above the soil. Martian soil was placed in a nutrient-rich solution and left to incubate for 12 days, after which the gases released from the soil were sampled by the chromatograph. It found that the gases released were consistent with biological activity metabolizing the nutrients. But a second sample was heated to sterilize it, and then analyzed. It yielded the same result as the initial sample, suggesting that nonbiological processes were at work. In other words, the gases were from chemical processes, not microbes.

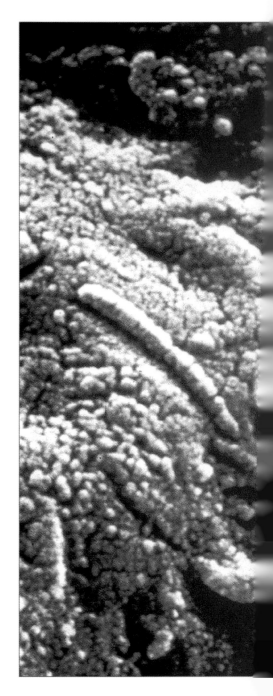

The Viking program, many researchers have concluded, has shown that Mars — or at least its surface — is sterile. Its thin atmosphere offers no protection from the Sun's harmful ultraviolet rays, which break down organic molecules and oxidize the soil. They contend that the labeled-release experiment recorded a chemical, nonbiological reaction with the oxides, not the metabolism of microbes. What's more, they note that oxides in the soil are toxic to living things. Proponents of the microbe interpretation counter this by pointing to tiny magnets attached to the Viking landers; they were covered with dusty material from the soil. If the soil were completely oxidized, then the magnets would not have attracted this material. The question remains unresolved to this day.

THE MARS ROCK

The Mars Rock — Allan Hills 84001 — mentioned earlier has been pivotal in giving previously unknown respectability to speculation about the existence of extraterrestrial life in outer space. While there are now many who are skeptical about the significance of the Mars Rock, since 1996 when President Clinton got involved in the debate, astrobiology has become *the* hot topic at international conferences. NASA's establishment of its own Astrobiology Institute to pursue the subject has given the discussion even more momentum.

There are three groups of rare meteorites — shergottites, nakhlites, and chassignites — which are collectively referred to as SNC (pronounced *snik*) meteorites. Only SNC meteorites contain concentrations of isotopes of noble gases — elements in the periodic table that tend not to bond with other elements. The concentrations of these isotopes are consistent with the composition of the Martian atmosphere, which confirms their Martian

The infamous organic shapes found inside the Martian meteorite AH84001 sparked a frenzied debate over the prospect of discovering life on Mars.

origin. Of the 34 known SNC meteorites in the world, just one — Allan Hills 84001, a shergottite meteorite — has become a lightning rod for controversy.

According to one theory, the story of this rock's odyssey began 4.5 billion years ago, when it was formed from molten rock, giving it the distinction of being one of the oldest objects in the solar system. Between 4 billion and 3.6 billion years ago, cracks in the rock became filled with carbonates and magnetite. Roughly 3.6 billion years ago, the rock was shattered by the impact from another large meteor. The shattered pieces of the original rock lay on the surface of Mars until a second, more recent, impact from another meteor launched a piece of the original rock into space where it drifted around the solar system for about 15 million years.

Approximately 13,000 years ago, it drifted into Earth's path. The rock entered Earth's atmosphere, lit up the night skies over Antarctica, and then thumped into the ground. It was soon buried under the glacial ice pack, where it was slowly churned inside until it emerged onto the icy surface roughly 500 years ago. In 1984, the rock was found by geologist Roberta Score and labeled AH84001 after the place where it was discovered, Allan Hills.

Somehow the meteorite was misidentified when it was shipped to NASA's meteorite processing lab at the Johnson Space Center in Houston, Texas, where it sat on a shelf for 8 years. Finally, geologist David W. Mittlefehldt analyzed a slice of AH84001. Because of the erroneous label, he thought it had come from an asteroid called 4 Vesta, but he determined that its chemical signature matched that of SNC meteorites instead.

On August 6, 1996, NASA scientist David McKay announced his findings. Taken together, they provided circumstantial evidence that AH84001 contained fossilized Martian microbes. There were four main pieces of evidence to support this claim.

SPECULATION

The astronauts visit one of the many robotic landers from other countries now conducting scientific experiments on Mars.

First, the rock contains amino acids and polycyclic aromatic hydrocarbons (PAHs), large molecules that result from the incomplete combustion of carbon fuels. They can be formed abiotically and have been found in comets and planetary nebulae, as well as in meteors. They are also found in the greasy residue left over from the decomposition of a dead organism.

Second, the rock has crystals of magnetite and iron sulfide that may be byproducts of microbes that feast on iron. Experts contend that such crystals cannot be produced abiotically.

Third, additional evidence is found in the mineralized shapes seen in the photograph of the rock. These shapes resemble that of a hypothesized nanobacterium. Discovered under a scanning electron microscope, these shapes — roughly 20 to 100 nanometers in diameter — simply look like bacteria.

Finally, still further evidence is in the age of the rock, which is consistent with the hypothetical wet Phyllosian period of Martian geological history.

At the time of the announcement, even McKay cautioned against fully embracing what seemed to be the obvious conclusion, that these were fossilized life forms from Mars. Since that day, doubts have mounted to the point where much of the scientific community now suspects that AH84001 was contaminated by microbes from Earth. Tests were conducted to examine the possibility of mishandling by the lab; after all, we know that the meteor was labeled incorrectly. Every possibility has to be examined.

Sadly, many have come to view David McKay as a latter-day Percival Lowell. It's not clear whether or not this is fair — that, in his case, the desire to find life on Mars overcame scientific discretion. The only thing certain is that AH84001, like the labeled-release experiment performed on *Viking*, is part of a larger puzzle about life on the fourth planet.

The latest rover missions have uncovered startling evidence of ancient shorelines, thus demonstrating that Mars perhaps once had oceans similar to those on Earth. But where did all that water go? Was there once life in those Martian oceans and, if so, what was its fate? There can be few more fascinating — or significant — questions facing science today.

If it is established that life arose independently on two or more worlds within our own solar system, the likelihood of finding life, perhaps even another civilization, around another star will be significantly increased. If, on the other hand, we find no indication of life on Mars, scientists will have to figure out why not — what makes conditions on Mars so different? Either finding would have profound implications.

The journey home will provide a lot of time for the crew of *Terra Nova* to contemplate the samples and data they've collected. In the event that they find evidence of life on Mars, the implications of this finding will cause all kinds of people in all kinds of professions to reconsider the Universe. Theologians may wonder anew about the complexity of creation. Politicians will consider how security is affected and what riches await them and their constituencies. Above all, scientists will delight in the possibility of new discoveries. Questions about the origins of life are part of a set of questions so large and so personal that every thinking person must surely address them.

The journey home begins as *Terra Nova* and *Gagarin* part company.

CHAPTER SEVEN

GOING HOME

"DAY WITH ITS BURDEN AND HEAT HAD DEPARTED, AND
TWILIGHT DESCENDING
BROUGHT BACK THE EVENING STAR TO THE SKY"
— Henry Wadsworth Longfellow, "Evangeline"

The crew's feelings as they prepare to leave Mars and begin the
long journey to Earth are mixed. They have worked hard but perhaps feel that their work is unfinished. They know that another
crew is already preparing to replace them. Some may feel the urge
to leave behind a mark or memento of their presence, as Charlie
Duke did when he left behind a photo of his family in a plastic
bag on the Moon. It will be surprising if the first men and women
on Mars do not feel strong emotions as their tour nears its end.

Terra Nova fires its main engines as it returns to Earth.

The last day on Mars. The crew has spent its final EVA buckling down the weather station and loading samples into *Gagarin*'s cargo bay. Each member of the crew takes a last look at the rusty hills that have been home for the past 60 days. The ruddy hues of Mars will soon be a memory.

The crew enters *Gagarin* and prepares for the return to orbit. This is the first time humans have ever launched from Mars. The sense of relief and accomplishment is palpable as they ignite the engines. With a flash *Gagarin* easily climbs back into the vermillion skies over Mars. The cameras aboard the two SEVs tilt upward and track *Gagarin*'s progress. Dust swirls around the rovers and through the base camp.

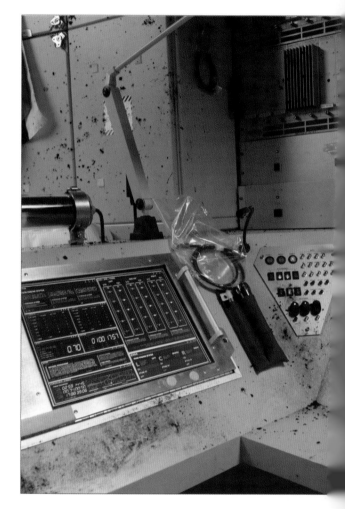

For a while, at least, there are again no humans on Mars. The crew's thoughts turn toward the long journey home. But nothing could have prepared them for what they find inside *Terra Nova*. A black, musty mold covers nearly every surface inside the spacecraft. At least, it seems like mold. It will have to be studied for its toxic qualities and, of course, scrubbed. Is it possible that, after surviving the extraordinary rigors of Mars, the crew will succumb to something as ordinary as mold?

Experience with space mold and other fungi dates back to the original *Mir* program. Cosmonauts began to notice that the view from a porthole on the orbiting space station was becoming degraded — a mold had taken root. They found mold on electrical contacts and polyurethane pieces; and parts of copper cables throughout the space station were oxidized as a result. As the fungi metabolize, they produce corrosive acids that can damage sensitive equipment. But space mold doesn't dine just on metal, glass, and plastic. It also regards skin epithelia, lipids, and other products of human activity as the equivalent of haute cuisine. Consequently, the air supply and the surfaces around the exercise equipment are loaded with mold feasting on the residue of human sweat and breath.

One hundred and seven species of fungi were found on *Mir*. The most common was *Penicillium chrysogenum*. It is not known for certain why, but a fungus in space spreads much more aggressively than it does on Earth. It seems that the increased exposure to radiation — 500 times more intense in space than on Earth — accelerates its rate of mutation. Suddenly and inexplicably, fungi in space can stop spreading and lie dormant, and then, just as capriciously, start growing again.

To fight mold, spacecraft before launch are filled with a gas that combines ethylene oxide and methyl chloride and is lethal to microorganisms. On the *ISS* and on long-duration flights, an antifungal cleanser is used regularly to wipe surfaces.

For the crew newly returned to *Terra Nova*, exiting Martian orbit takes priority over a fungus. The cargo module from *Gagarin* is transferred to its docking slot along *Terra Nova*'s truss. The return trip will take 6 months, much less time than the 11-month outbound trip. Mars and Earth are in constant motion relative to each other and have drawn toward an opposition. *Gagarin* is secured, with the crew areas placed in sleep mode, awaiting its rendezvous with the next visitors from Earth. The docking ring is unlocked, and the crew waves goodbye as *Terra Nova* gently pulls away. The two spacecraft cross the Martian terminator (the transition zone between night and day) for the last time. *Terra Nova* fires its main engine and begins the journey home.

When the crew return to the orbiting *Terra Nova*, they are confronted by a most unwelcome invader. As on *Mir* and the *ISS*, mold is a recurrent problem.

On board *Terra Nova*, the mold is everywhere. The crew spends many days "swabbing the decks," but they take the task in stride. Their mood is reflective, perhaps even reverent. The greatest journey in the history of mankind is now winding down.

August 31, 2031. Blue sky. Blue water. Seagulls utter raucous calls above the beach, a tropical breeze rustles the fronds of palm trees, and the blue-green Earth extends a warm embrace to six weary travelers. The *Orion* Earth-return capsule splashes down a few miles from the Kennedy Space Center in Florida. It is slightly off-target but soon tracked down and recovered. Private yachts and small craft of all kinds join the fleet as the aircraft carrier awaits offshore. Fireboats with their water cannons spraying in the air circle the whole procession. Horns toot and blow, and flags flutter in a surge of pride that transcends national boundaries.

The crew members of *Terra Nova* are a little wobbly as they climb from the hatch of *Orion*. One by one, they climb into the C-Horse 2 helicopter waiting to transport them away from the aircraft carrier and into a prescribed period of quarantine and to their first mission debriefing since returning to Earth. They are filled with wonder and joy to find themselves again on solid ground, at 1.0 G, the weight of the journey now lifted. In the few minutes since the *Orion* capsule returned to Earth, the crew has reveled in the contrast between the silent void of space and the noisy world teaming with life. They luxuriate in the warm, nurturing Sun, protected by a real atmosphere studded with clouds of water vapor that make fat summer rain. The crew has spent a lot of time contemplating the meaning of their accomplishments. Much has changed.

They are no longer citizens of one planet. They are now citizens of the solar system — one people, two worlds.

On December 22, 1968, during their return journey from the Moon, the crew of *Apollo 8* captured this image of Earth.

SHOCKING REPORT REVEALS
NAMES OF MARTIAN CONSPIRATORS

Washington, D.C. (filed by Dana Berry): Details of the imminent invasion of the planet Mars were disclosed today in a classified report leaked by a high-ranking government official. The report indicates that Martians have established a beachhead in Canada, at the offices of Madison Press Books in Toronto, and in the United States at the offices of Barron's Educational Series, Inc., in New York.

The names of individuals engaged in the front lines of this invasion have now been revealed. Jonathan Webb, regarded by many as a hero, deployed critical editorial skills when the Martians launched their literary weapons of mass confusion. On many occasions Webb successfully decoded cryptic Martian writings into a meaningful terrestrial language. Richard Lachman of QuickPlay Media was unfailingly helpful in coordinating a complex project. Together with Simone Garneau of Galafilm, he also provided an effective defense against repeated Martian attacks on accuracy. Even the Martians acknowledged the graphics genius of Gorette Costa, based at Costa Leclerc Design. The outstanding art-

work by Digital Dimension for Galafilm portrayed nearly every aspect of the invasion. Maria-Fernanda Barrios and Manon Barriault provided invaluable assistance. Wanda Nowakowska, now a veteran of two invasions from outer space, provided strategic oversight during the entire operation.

The report further commends the following individuals: Paul Lewis, President and General Manager of Discovery Channel Canada, whose vision made the project possible. At Madison Press, the names include Diana Sullada, Sandra Hall, Beth Martin, Susan Barrable and Oliver Salzmann.

The report also cites individuals who describe themselves as friends of the author, including Dan Durda, Ray Villard, Eric Chaisson, Keith Robinson, "Professor Andy" from "Other Times" Used Books, and the clamorous members of the IAAA. Their advice during the invasion proved critical. The report concludes with Olympus Mons-sized mountains of gratitude for the patience, help, and understanding of Rosa and Fima Kleinerman, Alla Savranskaia, and especially to that future denizen of the Red Planet, J.Z. Berry.

SELECTED BIBLIOGRAPHY

Bergreen, Laurence. *Voyage to Mars.* New York: Riverhead Books, 2000.

Boorstin, Daniel J. *The Discoverers.* New York: Random House, 1983.

Bradbury, Ray. *The Martian Chronicles.* New York: William Morrow, 1997.

Braun, Wernher von. *Project Mars.* Burlington: Collector's Guide Publishing, 2006.

Burrough, Bryan. *Dragonfly.* New York: HarperCollins, 1998.

Burroughs, Edgar Rice. *A Princess of Mars.* Toronto: Random House Canada, 1912.

Caiden, Martin, and Jay Barbree. *Destination Mars: In Art, Myth and Science.* New York: Penguin Studio, 1997.

Chaikin, Andrew. *A Man on the Moon.* New York: Viking Penguin, 1994.

Chaisson, Eric, and Steve McMillan. *Astronomy Today.* Upper Saddle River: Pearson Prentice Hall, 2005.

Clark, Arthur C. *The Sands of Mars.* New York: Harcourt Brace Jovanovich, 1952.

Couper, Heather, and Nigel Henbest. *Mars.* London: Headline Book Publishing, 2001.

Dupas, Alain. *Destination Mars.* Toronto: Firefly Books, 2004.

Ferris, Timothy. *Coming of Age in the Milky Way.* New York: William Morrow, 1988.

Gombrich, E. H. *Art and Illusion.* Washington, D.C.: Pantheon Books, 1960.

Hardy, David A. *Visions of Space.* New York: Gallery Books, 1990.

Hartmann, William K. *A Traveler's Guide to Mars.* New York: Workman Publishing, 2003.

Koestler, Arthur. *The Sleepwalkers.* New York: MacMillan, 1959.

Krupp, E.C. *Beyond the Blue Horizon.* New York: HarperCollins, 1991.

Lawrence, Richard Russell. *Space Exploration and Disasters.* London: Constable & Robinson, 2005.

Ley, Willy. *Watchers of the Skies.* New York: Viking Press, 1963.

Linenger, Jerry M. *Off the Planet.* New York: McGraw-Hill, 2000.

McCurdy, Howard E. *Space and the American Imagination.* Washington, D.C.: Smithsonian Institution, 1997.

McDougall, Walter A. *The Heavens and the Earth.* New York: Basic Books, 1985.

Morton, Oliver. *Mapping Mars.* New York: Picador, 2002.

Sagan, Carl. *Cosmos.* New York: Random House, 1980.

——. *Pale Blue Dot.* New York: Random House, 1994.

Sawyer, Kathy. *The Rock from Mars.* New York: Random House, 2006.

Schopf, J. William. *Cradle of Life.* Princeton: Princeton University Press, 1999.

Serviss, Garrett P. *Edison's Conquest of Mars.* Burlington: Apogee Books, 2005.

Shayler, David J. *Disasters and Accidents in Manned Spaceflight.* Chichester, UK: Springer Praxis, 2000.

Shepard, Alan, and Deke Slayton. *Moon Shot.* Atlanta: Turner Publishing, 1994.

Stoker, Carol, and Carter Emmart. *Strategies for Mars.* San Diego: American Astronomical Society, 1996.

Squyres, Steve. *Roving Mars.* New York: Hyperion, 2005.

Wachhorst, Wyn. *The Dream of Spaceflight.* New York: Basic Books, 2000.

Ward, Peter. *Life as We Do Not Know It.* New York: Viking Penguin, 2005.

Wells, H.G. *The War of the Worlds.* London: William Heinemann, 1898.

Wilford, John Noble. *Mars Beckons.* New York: Knopf, 1990.

Zubrin, Robert. *The Case for Mars.* New York: Touchstone, 1996.

——. *Entering Space.* New York: Penguin Putnam, 1999.

IMAGE CREDITS

Every effort has been made to correctly attribute all material reproduced in this book.
If any errors have unwittingly occurred, we will be happy to correct them in future editions.

DLR/FU	Deutsches Zentrum für Luft- und Raumfahrt
ESA	European Space Agency
Galafilm	Galafilm Productions (XII) Inc.
JHU-APL/NASA	Johns Hopkins University Applied Physics Laboratory/National Aeronautics and Space Administration
NASA	National Aeronautics and Space Administration
NASA/JPL/Caltech	National Aeronautics and Space Administration/Jet Propulsion Laboratory/California Institute of Technology
NASA/JPL/Malin	National Aeronautics and Space Administration/Jet Propulsion Laboratory/Malin Science Systems
NASA/JPL/Arizona	National Aeronautics and Space Administration/Jet Propulsion Laboratory/University of Arizona

INDEX

References to images are indicated by italicized page numbers.

A

A-9 rocket, 68
A-11 rocket, 68
Abwehr, 65
acidophiles, 160
Advanced Propulsion Lab (NASA), 15
aerobraking, 18
aero-descent heatshell, 27
aeroshell, 28, *28*
airbags, 83, 87
Allan Hills 84001 (meteorite) 83, 173–6, *173*
Alexandria, library of, 36
alkaliphiles, 160
allergies, 134, 135
Almagest, The (Ptolemy), 36
American Corporal rocket, 66, *66*
Ames Research Center, 108
amino acids, 159, 175
ammonia, 117, 159, 167
Angry Red Planet (film), 41
Antarctic expeditions, 21
Aphrodite, 32
Apollo missions, 128, 150
Apollo rovers, 131
Apollo 7, 11, *12*, 13
Apollo 11, 20
Apollo 17, 134, 135, 149
Applied Physics Laboratory (NASA), 122
areography, 54
Ares (Greek god), 54
Ares (rocket), 91, 92, 94, 96, *96*, 98
Ares Valley (Mars), 83
Aristarchus, 36
Aristotle, 34
Arctic expeditions, 21
Armstrong, Neil, 20, *21*, 128, 129
Artemia salina, 161
Ascraeus Mons (Mars), *49*
asteroids, 122, 174

Astrobiology (magazine), 83
Astrobiology Institute (NASA), 173
astronauts, *141, 144, 145, 148, 149, 151, 157, 158, 175*
  communications with, 28, 112–113, 128, 130
  daily routine of, 123, 130, 146, 148, 152
  and depressurization, 143
  educational requirements, 20
  and exercise, 113, *113*, 123
  group dynamics, 21, 22, 110, *111*
  hazards facing, 103, 104, 107, 119, 142–43, 145–46
  and hobbies, 111
  and illness, 111–13, 115, 135, 142, 143
  movies about, 41
  and nutrition, 116, 150–52
  personality profile, 20
  quarantine, 182
  and radiation, 114, 115, 145
  recruitment of, 20, 21
  and sleep deprivation, 111
  stresses on, 20, 109, 110
  training of, 22, 146
  and waste management, 117
  workload of, 110
Astronomia Nova (Kepler), 46
Atlantic Canali, 47, 54
Atlantis (habitat module), 26, 98, 99, *99*, 100, 114, *114*, 117, 131, 132, *133*, 135, 143, *144*, 150. (*See also* Trans hab)
atomic waste, 167
autoimmune deficiencies, 113

B

Baikonur Cosmodrome, 85, 98
Balloon Atmospheric Radar Survey Over Mars (BARSOOM), 132, 133, 137
Barker's Cave (Australia), 139
Barophiles (microorganisms), 160
Battlestar Galactica (film), 101
Bean, Alan, 131
Beer Sea (Mars), 54
Beer, Wilhelm, 53
belts, Van Allen, 75
Bibring, Jean-Pierre, 168
biomass, subterranean (Earth), 164
Bonestell, Chesley, 73
booster, 66, 68, 70, 82, 92, 94, 96, 118
Borowski, Stanley, 15
Botanical Production System (BPS), 108, 109
Bowie, David, 41
Bradbury, Ray, 150
Brahe, Tycho, 38–40, *38*, 44, 45
Brasier, Martin, 157
Brazilian Space Agency (AEB), 107
Brera Astronomical Observatory, 54
Bruno, Giordano, 38
Bulfinch, Thomas, 32
Bumper 8 (rocket), 66, *66*
Buran (shuttle), 98
Bush, George H.W., 80

C

California, University of, 83
Callisto (Jovian moon), 156
CAPCOM (spacecraft communicator), 92, 94, 96, 102, 108
Cameron, James, 83

Canadarm, 115
Canadian Space Agency (CSA), 21, 107
canali, 54, 55, 58
Cape Canaveral, 66, *66*
carbon dioxide, 59, 109, 163, 166
carbon 14 (C-14), 172
carcinogenesis, 115
Case for Mars, The (Zubrin), 101
Cassini, Giovanni, 47
Cassini Land (Mars), 54
Cassini probe, 120
Centauri Montes (Mars), 88
centrifugal force, 113
cerebral edema, 143
Challenger Deep (trench), 160
chassignites (meteorites), 173
chemolithoautotrophs (microbes), 163
chemoorganotrophs (microbes), 163
Chicago, University of, 159
C-Horse 2 (helicopter), 182
Chryse Planita (Mars), 83
Cimmerium (Mars), 54
circadian rhythm, 111, 112
Clinton, Bill, 80, 83, 173
Cold War, 107
Columbia (shuttle), 94, *95*, 102, 105
comets, 33, 34, 167, 175
communications, 82, 92, 94, 96, 98, 102, 103, 107–8, 128
Conrad, Pete, 131
corrosion on spacecraft, 120
Contact (film), 19
Copernic Continens (Mars), 54
Copernicus, Nicholas, 36–38
coronal mass ejection (CME), 115
Cornell University, 156
cosmonauts, 21, 22. (*See also* astronauts)
Cosmotheoros (Huygens), 47
creation, process of, 157–63

cretaceous-tertiary boundary (KT), 122
crew exploration vehicle (CEV), 100, 102
Curbean, Robert, 107, *107*
Cyanidium caldarium, 160
cyanobacteria, 157, 158, 163

D

Dao Valles (Mars), 22, *23*, 26, 30, 31, 102, 130, *130*, 135, 138, 140
Darwin, Charles, 55, 85
Darwinisim, 58
Da Vinci Code, The, 50
Deep Space 2, 79
Deinococcus radiodurans, 156
DNA, 115, 159
De Palma, Brian, 41
De Revolutionibus Orbium Coelestium (Copernicus), 37
Der Planet Mars, eine zweite Erde (Schmick), 55
Description de l'univers (Mallet), 34, *35*
Discovery (shuttle), *90*, 91, 105
docking, 24, *24*, *25*, 26, 102
dogs in space, 74
"Doom" (video game), 41
drogue parachutes, 27, *27*
Duke, Charley, 179
dust storms, 22, 53, 77, 78, 82, 131–33, 135–37, *135*, 145, *145*, 152
dysrhythmia, 112

E

Eagle Crater (Mars), 87, *87*
Earth
    as viewed from space, 15, *183*
    atmosphere of, 72, 118
    axial tilt, 52
    biochemistry of, 160–61, 163, 164
    and geocentric model, 34
    and glaciation, 138
    and gravity, 71, 139
    and heliocentric model, 34

and lava tubes, 139
–Mars comparisons, 52, 109, 112, 117, 133, 138–39, 143, 145, 150, 152, 159, 163–64, 166, 169, 176, 182
and meteorites, 174–75
orbit of, 45, 52, 82, 98
origin of life on, 83–84, 157–58, 162–63
prime meridian of, 53
radiation on, 145
subterranean biomass, 164
Earth return capsule (ERC), 19, 100, 182
ecopoiesis, 167
egress platform (*Terra Nova*), *124*, 125, 132
Edward the Confessor, 33
Eisenhower, Dwight, 71, 74
Einstein, Albert, 45
electromagnetism, 107, 108
Elektron unit, 104
Elysium (Mars), 54
E-mail, 108
Enceladus (moon of Saturn), 156
Energia (rocket), 98, 101
eras, geological (Mars), 168–69
Erebus Crater, *82–83*
Europa (Jovian moon), 156
European Space Agency (ESA), 21, 77, 85, 107, 168
epicycles, 45
EVA suits, *128*, 129, *129*, 146, 189
exercise, 113, *113*, 123
experiments, 50, 64, 72, 108, 123, 132, 137, 139, 146, 148, 151–52, *151*, 158–59, 158, 162, 172–73
Explorer I, 75
extra vehicular activity (EVA), 104, 115, 136, *136*, 137, *137*. (*See also* EVA suits)
extremophiles, 156, 160, 166, 169

F

Ferroplasma acidiphilum, 160
fire, 116–17
Flammarion, Camille, 58
Florida, space center. (*See* Kennedy Space Center)
Fleury, Robert, 50
flyby, 120, *120–21*
Flyer, 79
Foale, Michael, 109, 111
Fogg, Martyn, 167
433 Eros (asteroid), 122, *122*
4 Vesta (asteroid), 174
Frederick II, king of Denmark, 38
free radicals, 125

G

G-1 rocket, 70
G-4 rocket, 70
Gagarin (MADV), 18, 19, *19*, 22, 24–31, *26*, *27*, *28*, 98, *124*, 125, 128, 130, 143, 146, *177*, 180, 181
Gagarin, Yuri, 74
Gaia Hypothesis, 166
Galalea, 32
Galilei, Galileo, 38, 44, 46, *51*
Galileo probe, 120
Galileo's code, 50
gamma ray bursts (GRB), 115
gas chromatograph and mass spectrometer (GCMS), 172
gas exchange experiment (GEX), 172
gas leaks, 103–4, 107
genome, human, 158
George III, king of England, 52
Georgium Sidus (star), 52
Gibbons, Jack, 83
glaciation, 138
Glenn Research Center (NASA), 15
global warming, 166
glycine, 159
god of war (Mars), 34, 63
Goddard, Robert H., 64, *65*
Goddard Space Flight Center (NASA), 65

Goldilocks Zone, 156
Goldin, Daniel, 83, 84
"Good Day Sunshine," 112
Gorodomlya Island (Russia), 70
Graves, Robert, 32
gravity
    artificial, 18, 113, 119
    boost, 15, 119, 120
    micro, 108
    on Earth, 139
    on Mars, 77, 113, 130, 147
gravity-assist maneuvers, 84, 119, 120
greenhouse gases, 166–67
Groettrup, Helmut, 69–70
Groettrup, Irmgard, 70
Group for Investigation of Reactive Movement (GIRD), 67
Guggenheim, Daniel, 65
Gulliver's Travels (Swift), 47
Gusev Crater (Mars), 85, 126, 169, *170*, *171*
Gusev, Matvei, 85

H

Hadriaca Patera (Mars), 26, 139, 140
Hall, Asaph, 51
Halley, Sir Edmund, 33
Halley's Comet, 32–34, *33*
halophiles, 160
Harold, king of England, 32, 33, *33*
Hayabusa (spacecraft), 122
hazards, spaceflight, 103–4, 107
Hellas Basin (Mars), 21, 102, 133, 169
Hellas Planitia (Mars), 152, 153
Hells Pond (MA), 64
hematite, 85
Hershel, William, 52, *52*, 53
Hesperia (Mars), 54
Hoffman, Jeffrey, 91
Holocaust, 71
Holst, Gustav, 41, 150
Hubble Space Telescope, 28, 105, 117, 118

human genome, 158
Huygens, Christiaan, 47
hydrogen, 15, 104, 115, 159, 169
hydrogen peroxide, 134
hydrothermal vents, 162
hyperthermophiles, 156, 161, 166, 169

I

Iapetus (moon of Saturn), 156
ICBM (intercontinental ballistic missile), 70
Indiana, University of, 163
International Space Station (ISS), 21, 22, 28, 70, 103–5, *105*, 107–8, *107*, 111–12, 114–15, 117, 181
instant messaging, 108
Institut d'astrophysique spatiale, 168
Institute for Genomic Research, 158
Invaders from Mars (film), 41
Iowa, University of, 78
iron sulfide, 175
iron-sulphur world theory, 162
Is Mars Habitable? (Wallace), 61

J

James VI, king of Scotland, 39
Japanese Aerospace Exploration Agency (JAXA), 21, 107, 122
jet-packs, 129
Jet Propulsion Laboratory (NASA), 15, 79
John, Elton, 41
Johns Hopkins Hospital, 65
Johnson Space Center, 14, 83, 103, 113, 174
Johnston Island, 66
Jupiter, 34, 44, 105
Jupiter C rocket, 71, 74

K

KT (asteroid), 122
Kapustin Yar (firing range), 70

Katyusha rocket, 67
Keck, W. M., 169
Kennedy, John F., 74, *75*
Kennedy Space Center, 91, 92, 182
Kepler, Johannes, 37, *37*, 39, 44, 45, 50, 51
Kilmer, Val, 41
Kordev, Sergei, 70

L

labeled release experiment, 172
Laika (dog in space), 71
landers, 27, 77–80, *80*, 84, 85, 172, 173, 176
La Planete Mars (Flammarion), 58
Laputians, 47
latitude, 53
lava tubes, 139, 142
Lawrence Livermore National Laboratory, 97
laws of planetary motion, 46
Leningrad Gas Dynamic Laboratory (GDL), 67
Leonov, Alexei, 74
library, Alexandria, 36
Lindbergh, Charles, 65
Linenger, Jerry, 116
Lippershey, Jan, 46
lithium perchlorate canister, 116
lithoautotrophs, 161
Longfellow, Henry Wadsworth, 179
longitude, 53
Longomontanus, Christen Sørensen, 46
Lost in Space (TV), 41
Lovelock, James, 166
Lowell Observatory, 58
Lowell, Percival, 58, 59, *60*, 61, 64, 67, 155, 176
Lucid, Shannon, 111

M

McArthur, Bill, 112
McCartney, Paul, 109
McKay, Chris, 166, 167
McKay, David, 83, 174, 175, 176

macro molecules, 159
Madler, Johann, 53
MADV (Mars ascent/descent vehicle), 18, 22
magnetite, 174, 175
Mallet, Alain Manesson, 34
Man of Mars (book series), 41
Maraldi, Giacomo, 52
Marianas Trench, 160
Mariner IV, 77
Mariner VI, 77
Mariner VII, 77
Mariner IX, 48, 49, 77, 78
Mars, *10, 24, 25*
    atmosphere on, 52, 53, 143, 145, 166, 167, 173
    axial tilt, 52, 53, 166
    biochemical origins of, 162–63
    and dust storms, 133, 134, *135*, 145
    distance from Earth, 53
    early telescopic observations of, 46, 47
    –Earth comparisons, 52, 109, 112, 117, 133, 138, 139, 145, 150, 152, 159, 166, 169, 176, 182
    face of, *48, 49, 49*
    geological history of, 164, 168–70, 175
    latitudes/longitudes, 53
    and lava tubes, 139, 140
    legends and myths about, 41, 50, 79
    length of day on, 47, 138, 166
    length of trip to/from, 22, 181
    life on, 58, 59, 61, 168, 169, 172, 175, 176. (*See also* microbes)
    map of, *56, 57*
    moons of, 50, 76, 77, 80
    movies about, 41
    opposition (planetary) of, 58, 107, 119, 181
    orbit of, 52, 53
    and parallax, 44

prime meridian, 53
regolith, 134, 166, 167
sols of, 152
temperatures on, 166
    (*See also* water)
Mars, god of war, 34, 63
Mars (Lowell), 58
Mars 2, 77
Mars 3, 77, 78, 79
Mars, a World Engaged in the Struggle for Survival (Dross), 55
Mars and Its Canals (Lowell), 58, 61
Mars ascent/descent vehicle. (*See MADV*)
Mars As the Abode of Life (Lowell), 58
Mars Attack (film), 44
Mars Climate Orbiter (NASA), 79, 84
Mars Design Reference (MDR), 101
Mars Exploration Rover-A (MER-A), 85
Mars Express Orbiter, 77, 85, 168
Mars Global Surveyor (satellite), 59, 82, 83, 84, 88, *89*, 135, 152, 169
Mars Odyssey Orbiter (NASA), 169
Mars Observer (satellite), 82
Mars Pathfinder, 83, 84
Mars Polar Lander, 79, 84
Mars Reconnaissance Orbiter, 28, 80
"Mars, the Bringer of Wars" (Holst), 150
Martian Chronicles, The (Bradbury), 149
martians, *42–3*, 59, 61, 64
Mawrth Vallis (Mars), 28, *29*
Medicean Moons (Jupiter), 47, 51
meteorites, 83, 173, *173*, 174, 175, 176. (*See also* specific names)
Methanococcus jannaschii, 161
Mercury (planet), 51
Mercury (mission), 110

Metharme, 32
methane, 169
microbes, 160–61
Micromegas (Voltaire), 50
microwave radiation, 114, 145
midnight sun, 111
Miller, Stanley, 159
Mir, 21, 22, 109, 112, 116, 117, 180, 181
Mission to Mars (film), 41
Mittelbau-Dora (factory), 68
Mittlefehldt, David W., 174
mold, 180, *181,* 182
Moon (Earth), 13, 22, 46, 51, 53, 64, 74, 80, 84, 94, 105, 157, 179
     gravity on, 148
     mission to, 91, 148
     orbit of, 119
moons
     of Mars, 50, 76, 77, 80
     of Jupiter, 47, 51, 156
     of Uranus, 53
     of Saturn, 53, 156
Mount St. Helens (volcano), 135
Musgrave, Storey, 109
My Favorite Martian (TV), 41
Mysterium Cosmographicum (Kepler), 44

N
nakhlites (meteorites), 173
napalm, 68
NASA (National Aeronautics and Space Administration), 20, 74, 88, 107, 113
National Air and Space Museum, 79
National Research Council, 142
near-Earth asteroids (NEA), 122
Nebelwerferi (launcher), 68
Neff, Donald, 79
Newton Ocean (Mars), 54
New York Times, The, 40
Nicholas V, pope, 36
nitrogen, 117, 145, 167
Nowak, Lisa, 110

Nozomi (spacecraft), 79, 84
Nuclear Engine for Rocket Vehicle Applications (NERVA), 15
Nuclear Thermal Rocket (NTR), 15, 18, 100

O
obstacle avoidance system, 83
occhiali, 46
oligotroph, 161
Olympus Mons (volcano), 78, *78*
1036 Ganymede (asteroid), 122
Opportunity (rover), 63, 84, 87, *87,* 163, 168
osteoporosis, 112
orbital insertion, 14, 18, 82, 84
orbital rendezvous, 24, *24, 25*
orbiter, 77, 80
Orion (CEV), 102
Orion (ERV), 182
Osiander, Andreas, 37, 38
Ovid, 31
Oxford, University of, 157
oxygen, 15, 64, 65, 68, 104, 109, 116, 117, 163
ozone, 166

P
Padua, University of, 46
Pal, George, 41
Paphus, 32
parachutes, 27, *27,* 28, *28,* 66
parallax, 44
Pavonis Mons (Mars), 49
Penicillium chrysogenum, 181
perfluorocarbons (PFCs), 167
perihelion, 133
Phobus, *76,* 77, 80
phosphate, 167
Phyllosian era, 168, 175
Pickering, William H., 58, 59
Piezophiles, 160
planetarians, 47
planetary motion, laws of, 46
planetary opposition, 54

Planet Mars, The: A Second Earth (Schmidt), 55
Planet Suite, The (Holst), 41, 150
Plato, 37
platonic solids, 44
polar icecaps, 52, 53, 59, *59,* 66, 88, *88,* 161, 166
polyethylene, 115
Polyakov, Valery, 113
polycyclic aromatic hydrocarbons (PAHs), 175
Polyextremophile, 161
Polyus (antisatellite weapon), 98
potassium, 167
potassium hydroxide, 103
Pratt, Lisa, 163
prime meridian, 53
prokaryotic cells, 158
Progress Supply Module, 22
Prometheus test bed probe, 15
Psychrobacter cryohalolentis, 161
Psychrophiles, 161
Ptolemy, 36
Pygmalion, 31–33
Pythagoras, 36
Pythagorean school, 36
pythagorean solids, 44

Q
quarantine, 182
Quest (airlock), 104

R
R-3 rocket, 70
R-7 rocket, 70
Race to Mars (TV), 22, 145
radiation, 18, 20, 74, 113–15, 119, 123, 125, 134, 145, 156, 161, 166, 181
     shielding from, 114, *114,* 125, 145, 166
radioresistants, 161
Reagan, Ronald, 107
Red Planet (film), 41
Red Planet, the, 13, 14, *24, 25,* 33, 34, 53, 63, 79, *80,* 83, 102–3, 164

Redstone (rocket), 71, 74
Refractor telescope, 47
regolith, 134, 167
remote manipulation arm, 115, *115*
retinal hemorrhaging, 145
retrograde motion, 34
Rheticus, Georg Joachim, 37
rift valley, 164
RNA, 159
Robert-Fleury, Joseph-Nicolas, 50
Robinson Crusoe on Mars (film), 41
rockets
     ballistic, 67
     clubs for, 65, 67
     ICBM, 70
     liquid-fuel, 64, 65, 67
     nuclear thermal (NTR), 15
     proton, 82
     retro, 27
     ring, 26
     robot, 18
     solid fuel, 67
     (*See also* specific names)
"Rocket Man" (John), 41
Roswell (NM), 65
rovers, 63, 83, 84, 85, 87, *87,* 163, 168, 169, *171*
Russian Space Agency (RKA), 107

S
Sagan, Carl, 41, 61, 77, 166, 167
"Sailor Moon" (cartoon), 41
Saturn, 34, 44, 50, 120
Saturn V Apollo rocket, 15, 92, *93,* 96, *96*
Schiaparelli, Giovanni, 54, 55, *56, 57,* 61
Schmick, Professor, 55
Schmitt, Harrison "Jack," 134, 135
Schopff, William, 83, 157
Score, Roberta, 174
Secchi Continens (Mars), 54
Secchi, Father Angelo, 54
Shackleton, Ernest Henry, 11
shergottites (meteorites), 173

Shewanella benthica, 160
Shirase A rocket, 94, 96–98, *99*
shuttles, 66, 91, 94, *95*, *96*, 98
Siderikian era, 168
Silence of the Lambs (film), 110
Skylab IV, 110
SNC (meteorites, group of), 173, 174
sodium perborate, 134
sodium hydroxide, 134
Sojourner (rover), 83, 84
solar max, 18, 19
solar radiation, 18
Soyuz (spacecraft), 103, 108, 116
spacecraft, 24, *24*, *25*, 26
communicator. (*See* CAPCOM)
and mold, 180, *180*, 181, 182
Space Exploration Initiative, 80
space race, 41, 71, 74, 82
spin-induced artificial gravity, 113
Spielberg, Steven, 42
Sputnik I, 68, 69, *69*, 74
Sputnik II, 71
Sophocles, 155
"Spiders from Mars" (Bowie), 41, 150
Spirit (rover), 85, 169, *171*
Squyres, Steven, 156
SSTAR (small, sealed, transportable, autonomous reactor), 97
Star Trek (TV), 41
Star Wars (platform), 98
Stehling, Kurt, 75
stromatolites, 157
subterranean biomass, 164
Sulfolobales solfataricus, 160
Sun, 18, 34, 36, 105, 133
sunspots, 18, 19
superior conjunction, 107
surface exploration vehicles (SEV), *97*, 131, 180
Swift (satellite), 115
Swift, Jonathan, 47, 51

synodic period, 34
Syrtis Major (Atlantic Canali), 47, 54

T
taches blanches, 52
telescope, 46, *46*, 47, 53, 169
temperature, 132–33, 142, 146, 156, 160–61, 162, 166, 172
Tereshkova, Valentina, 75
terraforming, 166–67
Terra Nova (spacecraft)
and artificial gravity, 113, 119
components of, 14, *16*, *17*, 19, 100
corrosion on, 120
docking with, 102
fire on, 116–17, *116*
flight of, 18, 19, 100, 108, 119, 120, 176
and *Gargarin*, 22, *24*, *25*, *177*
and mold, 180–82
new technology on, 15
shielding of, 114, *114*
speed of, 123
Trans Hab, 16, 17, *16*, *17*, 100, 132, *132*. (*See also Atlantis* habitat module)
flyby (Venus), 120, *121*
Terra Sirenum (Mars), 80, *81*, 88
Tharsis Montes (volcanoes), *48*, *49*
Tharsis Bulge (Mars), 164, *164*
Theiikian era, 168
thermophile (organism), 161, 164
Thurston Lava Tube (Hawaii), 139
time capsule, 139
Time magazine, 79
Tokarev, Valery, 112
Truth, Sojourner, 83
tungsten, 15
25143 Itokawa (asteroid), 122
Twister (film), 132

Tyrrhena (Mars), 54

U
undersea exploration, 21
Universe
geocentric model, 34, 36
heliocentric model, 36
Ptolemic model, 37
and Pythagorean School, 36, 37
Copernican model, 36–38, 45
theory of (Brahe), 39, 40
Urey, Harold, 159
Urania (muse), 38
Uraniborg, 38, 39, 44, 52
uranium carbide, 15
uranium oxide, 15
Uranus, 52, 53
Utopia Planitia (Mars), 80
UV (ultraviolet) radiation, 145, 166

V
V-1 rocket, 68
V-2 rocket, 41, 66, *66*, 67–70, 101
Valles Marineris (Mars), 26, 49, 164, 169
Van Allen, James, 74
Vanguard rocket, 71, *71*
Vastitas Borealis (Mars), 163, 165
Vehicle Assembly Building, 91, 92, *92*
Venter, Craig, 158
Venus, 15, 51, 102, 105, 119, 120, 121
Verein für Raumschiffahrt (rocket club), 65
Vergeltungswaffe 2 rocket, 67
Victoria Crater (Mars), *62*, 63
Viking (lander), 79, 80, *80*, 172, 173, 176
volcanoes, 26, 48, 49, 78, 135, 139, 142
Voltaire, 50, 51
von Braun, Wernher, 68–69, *69*, 72–73, 74, *75*, 101
Vulkan rocket, 101

W
Wächtershäuser, Günter, 162
Wallace, Alfred Russell, 61
Walz, Carl, 111
War of the Worlds, The (book), 40, 55, 63, 65
War of the Worlds, The (films), 41, 42, *42*, 43, *43*
waste management on spacecraft, 117, 118
non recyclables, 118
water, 18, 52, 54, 83, 156, 159, 182
and organisms, 160–61
and origins of life, 162–63
on Mars, 59, *59*, 61, 82, 85, 104, 109, 117, 138, 139, 146, 152, 156, *157*, 163, *165*, 167, 168, 169, 176
weightlessness, 18, 112
Welles, Orson, 40, *41*
Wells, H.G., 40, 55, 63, 64, 65
When Worlds Collide (film), 41
White Sands Missile Range, 66
white spots, 52
W. M. Keck Telescope, 169
Wright brothers, 64, 79

X
Xanthe Terra (Mars), 49
xerophile, 161
X-rays, 114, 145

Y
Yost, Edna, 64

Z
Zubrin, Robert, 101, 167
Zvezda (module), 104

Manuscript Editor/Project Editor
JONATHAN WEBB

Book Design
COSTA LECLERC DESIGN

Editorial Director
WANDA NOWAKOWSKA

Art Director
DIANA SULLADA

Production Manager
SANDRA L. HALL

Printed by
SNP LEEFUNG, CHINA

RACE TO MARS
was produced by
MADISON PRESS BOOKS